Praise

WISDO[illegible]

"This is good stuff for the busy modern life: practical, simple, and wise."

—Jack Kornfield, author of
A Path with Heart

"Soren Gordhamer's brave and smart book on how to stay connected (i.e., mindful) while being electronically connected is a must for those whose lives tend to be on the virtual side. He is not advocating a backward step away from technology but an inward step to a new relationship with technology that is creative and liberating."

—Joan Halifax, Abbot,
Upaya Zen Center

"This may well be the first book of its kind. A book of rapid development and the rewriting of long conditioned programs. This is liberation by wrapped attention to the mind-screen and the comings and goings there-on."

—Stephen Levine, author of
Turning Toward the Mystery

"This is the instruction manual that should come with our iPhones and Blackberries . . . teaching us how to put them down."

—Chris Sacca,
former Google Head of
Special Initiatives

"*Wisdom 2.0* is a pragmatic, creative, and fun guide to applying ancient insights to our everyday technology-rich lives. This book reveals how to utilize ordinary circumstance for real transformation."

—Sharon Salzberg,
author of *Lovingkindness:*
The Revolutionary
Art of Happiness

"Utterly buried in the world of Twitter, blogging, email, online social networks, cell phones, text messages? *Wisdom 2.0* offers readers a ray of light, a breath of fresh air and tranquility for the constantly connected and eternally wired."

—Mark Grimes, founder
of Ned.com and Better World
Media Network

WISDOM 2.0

Ancient
Secrets
for the
Creative
and
Constantly
Connected

WISDOM 2.0

SOREN GORDHAMER

HarperOne
An Imprint of HarperCollins*Publishers*

HarperCollins books may be purchased for educational, business, or sales promotional use. For information please write: Special Markets Department, HarperCollins Publishers, 10 East 53rd Street, New York, NY 10022.

HarperCollins Web site: http://www.harpercollins.com
HarperCollins®, ®, and HarperOne™ are trademarks of HarperCollins Publishers.

Book design by Ralph Fowler / rlf design

FIRST EDITION

Library of Congress Cataloging-in-Publication Data is available.
ISBN 978-0-06-165151-9

09 10 11 12 CW 10 9 8 7 6 5 4 3 2 1

Contents

PART 3

The Browser: Creative Mind

Introduction: Establishing a Connection

Your Profile

Checking to see if we have a secure connection....
This will only take a few minutes.

Do you really think you have the time to be reading this book? I mean, consider all the other activities you could be doing: IM-ing, e-mailing, or texting a friend; updating your profile page on a social network like MySpace, Facebook, or LinkedIn; viewing the latest news on Yahoo! or the user-directed news site Digg.com; playing a game online; calling someone on your cell; creating or watching a video on YouTube; listening to your iPod; microblogging on Twitter; or exploring the Web via the browser plug-in StumbleUpon.

With so many options and considering how crazy busy you are, you are no doubt questioning if you have the time for this book. In fact, as you read these words, there is likely a communication device within your hand's reach, ready and waiting to be used at the first moment of boredom. To be worth your valuable time, this book better be damn good, and I must know who you are and what you seek. I have a tough job indeed, but if you give me a handful of paragraphs, I will do my best to address these.

Who are you? You are passionate, networked, busy, and ambitious. You crave cool gadgets that allow you to communicate in new ways, and you want to stay up-to-date on the most innovative Web sites and Internet tools. In doing so, you are almost always connected through a communication device, be it a Blackberry, cell phone, or computer. You thrive on information and want to be a creator, not just a consumer, of it. While you read news and watch videos, you also develop your own content through text, photos, and video. In all your efforts, you seek to accomplish as much as possible, and you get frustrated at anything that gets in your way.

You value honesty, and despise insincerity and showiness. While you could never see yourself living like the Dalai Lama or the pope, and you cringe at ooey-gooey niceness, you appreciate genuine kindness and thoughtfulness. Even a simple act—a note or gift from a friend—warms your heart. You see yourself on the side of the good, and you want to have a positive impact on the world.

You seek to live healthy by eating right and taking care of yourself, but you find doing so hard amidst such a busy life. You try to rid yourself of your daily espresso habit (or at least reduce it from a double three times to a single once a day) but you believe that with such an active life, you need the extra boost. You consider and, at times, join a yoga or exercise class, but while there you cannot help thinking about who might be trying to e-mail or call you while you are "disconnected." You check your phone and e-mail messages as soon as class is over, and you wonder why your instructor will not let you bring your Blackberry into class. You think of ways you might sneak it in. To be away from your communication devices for very long feels daunting.

You tell yourself that you must be constantly connected because of family, work, or school. "What if a friend, coworker, or family member needs to get ahold of me right away?" you say to yourself. "I must be reachable all the time." But you sense that there is more to your desire than this, as you can think of few examples when it was absolutely vital that you received information seconds after it was sent.

Though you greatly benefit from technology, you also notice some of the negative effects of such constant contact with it: You feel hurried and anxious more of the time; you eat faster and often do so while looking at a computer screen; you spend less time outside; you are not able to focus on one thing for very long; and you often experience a steady mental weariness. You know that stress can increase your chance of burnout and depression and can give you a higher risk of heart attack and stroke, but you do not identify as a stressed person. You are simply trying to live as creatively and effectively as possible. If you could get just as much done without feeling hurried and overwhelmed, that would be great, but you are not sure how to do so.

While you would never dream of giving up your cell phone, iPod, e-mail, or social network, you secretly question the impact they have on your life. You cannot quite explain it, but amidst your fast-paced, networked life you have an underlying sense that something is missing.

OK, so how did I do? If I am right (or pretty darn close), read on . . .

The Something That Is Missing

Status: Connected. Signal: Very good.

Still with me? Good. Now, you may be wondering, *how does he know so much about me?* You do not likely wear a twenty-four-hour webcam so I can see your every move. The answer is that I am just like you. I too live immersed in the technologies of our age.[1] Of course, you and I may have some differences. You may be more networked through one-to-one communication via cell, e-mail, IM, or text message; I may do so more through social networks like Facebook or online collaborative environments such as Wikipedia. However, we both seek to live with less stress and more effectiveness while networked.

So we are similar, I can hear you thinking, *but could you please get to the point?*

Of course. I know you are extremely busy, so let's get to it. Besides, I have told you that something is missing in our attempt to live creatively and make a positive impact on the world, but I have not yet informed you what this missing something is—and we are already on page 5! So let's move forward.

The something missing is an attention to the quality of our mind. Let me explain. A clear, focused mind is essential to any creative effort. Technology alone, however, does not guarantee us access to this state. In fact, with little focus of mind, the technologies meant to help us with a task can have just the opposite effect: The tools that are supposed to save us time can take more of it. Those created to help us communicate can actually lead to less genuine purposeful communication. And those meant to enhance our creativity can diminish it. This is why we can spend ten hours on our computer and find ourselves thinking afterwards, *Just what did I accomplish in all that time?*

The world has become ever more connected as over 1.4 billion of us have Internet connection[2] and roughly half the world population now has cell phones, but as our use of these devices has expanded and increased, so too has our stress.[3] In one study, three-fourths of participants reported that they have more on-the-job stress than a generation ago.[4] In another international study, two out of three respondents associated information overload with loss of job satisfaction; 42 percent attributed ill health to this stress; 61 percent said they have to cancel social activities as a result of information overload; and 60 percent claimed they are frequently too tired for leisure activities.[5] While it is amazing how connected we can live, if we are not careful, such a life can increase our stress and hinder our progress on the efforts that truly matter to us.

The something missing is not more tools or technologies, but the state of our consciousness. It is the lack of connection to the place inside us of ease and focus, what I refer to in this book as *the creative* or *creative mind*. This is the mind that experiences life freshly, thinks out of the box, finds new solutions to old problems, and expresses ourselves in innovative ways. It allows us to write a report or term paper with clarity and ease in three hours instead of banging our head against a wall and getting it done in eight.

When we do not see the importance of our mental state, we can be incredibly technologically savvy with the most recent iPhone, the snazziest computer, and have thousands of friends on the newest, hippest social network, but we relate to all these with the same old mind. We use cutting-edge technologies with the same mental staleness with which we related to the old ones. While technologies change and improve, our mental framework stays the same. Thus, we have the latest technologies, but the same old mind!

This old mind tries to impress people with its knowledge, goes through the day seeking pleasant experiences and avoiding unpleasant ones, dreads possible failure, checks messages every few minutes hoping for good news, worries what other people will think, feels lonely when not online, and continually compares itself with others. Essentially, it always looks for something out there to satisfy it. In this state, the moment we are living right now is never enough; there is always a future one that we believe will bring us the satisfaction we seek. In the state of old mind, it does not matter how fast our Internet connection, how sophisticated our hand-held device, or how well-paying our job . . . life is dull.

The frustration we feel comes from not making progress in living the potential we know is possible when we are free of old mind. To do this, we must be as knowledgeable about our internal world as we are about the external, as attuned to our inner technologies as to the outer ones. The art of such a life is the subject of this book. Rather than a scientific or philosophical exploration of this issue, it focuses on practical approaches and tools that can be performed in our daily lives. At the end of each chapter, I include an Apply to Life section focused on putting the lesson of that chapter into practice. Throughout the book, I also include daily practices to incorporate into your life.

Though years from now some of us may renounce technology and return to the ways of the Luddites, for most of us our interaction with it will only deepen, making us ever more connected in the years ahead. The question for most of us is not *if* we will use and be connected through technology, but *how* to do so in a way that enhances our creativity and well-being instead of adding to our frustration and stress. This book charts a map on this journey. It illustrates how the qualities that add stress to our lives are the same ones that limit our access to the creative mind, how what is best for our own well-being is also what is most helpful in our creative work. Therefore, we do not have to choose between effectiveness and calm. We can have them both. The choice is really between creativity/ease and staleness/anxiety.

I wrote this book not as an outside observer trying to study the population of those constantly connected. This is much more personal. It instead emerged from a wake-up call I had after several years totally immersed in technology, particularly the online world, where my desire to be connected all the time took precedence over just about every other facet of my life, including my mental and physical well-being. The tools that were once my humble servants and I enjoyed had become my overbearing masters, directing much of my life. I realized that if I continued my constant late nights surfing online, religiously checking my e-mail every few minutes, visiting every cool new Web site, regularly clocking twelve-hour days, and never swaying far from a communication device, my health and well-being would continue to deteriorate. As such, I was not sleeping well, my relationships were in chaos, and I had less free time. Though this was certainly concerning, what really floored me was what little progress I was making in such a life. I had more stress, was completing less of what was important to me, and was not making a positive contribution to the world as far as I could see. It was a lose-lose-lose situation. I sensed that there had to be another way.

I was not always such a rabid techie; in fact, I'd been far from it. I had previously spent a year on a global environmental walk, trekking for six months across the United States, three months through Pakistan, and another three months across Japan, all without a cell, computer, or often a map. For several years, I ran a Bronx-based nonprofit to teach stress-reduction programs to incarcerated teens in New York City. For much of my life, a techie I was not, but once I turned my attention to the online world, I became hooked, big time. I found, however, that while I had taken on some challenging projects in my life, finding a healthy balance with the connected world was as big as any I had previously attempted.

In this realization at just how much various technologies ruled my life, I was not (and am still not) willing to renounce my interest in or use of them. However, I am also not willing to let them rule my life in the way they did. I sought a middle way, a means to use the great technologies of our age creatively and effectively instead of habitually. I wondered where to look for guidance in this effort and discovered that ancient wisdom traditions such as Zen had much to offer on the subject. To tackle this formidable task, I realized that I had to merge my two interest areas: the path of technology and the path of wisdom. Out of that exploration, this book emerged. It offers tools that I have by no means mastered, but have found great use in implementing and practicing.

This book is not the four steps to retiring young (I haven't), or how to become a part of the new Internet rich (I am not), or how to guarantee that you rule the Web (I don't). I have nothing against these efforts, and some material here may even help achieve these goals, but the focus here is on living with deeper connection and greater ease; it's about creatively instead of stressfully responding to the conditions in our lives. I have met enough multimillionaires to know that you can be as miserable on a private jet as you can on a crowded city bus, and have known enough techies to learn that you can be just as lonely with thousands of friends on a social network as you can with only Tom on MySpace. Something is underlying the conditions in our lives, determining the level of our suffering or ease. To know this is the path of wisdom. To live it in our day and age while benefiting from the technologies of our time is Wisdom 2.0.

PART 1

THE POWER SOURCE: CONSCIOUSNESS

A teacher of mine used to say, "Do what you normally do; just do it with awareness." He meant that if you bring consciousness to any activity, your experience of it changes. If you make conscious a conversation, you better know what to say and can more fully listen to the other person; if you make conscious your work on a project, you can more easily see what it needs to progress; if you make conscious an unskillful habit, you can better understand why you follow it and how to release it. If you add consciousness to any activity, the nature of it changes, and the creative is illuminated.

The same is true in our use of technology. For it to be a creative force in our lives, something we skillfully use rather than feel used by, our first task is to make it conscious. Because consciousness is the foundation in the path of wisdom, in Wisdom 2.0 consciousness, especially in how we relate to technology, is key. Though this is addressed throughout the book, the following tools more directly highlight the key elements in this art.

Staying Charged

Even the best technologies fail without an adequate power source. That we are connected may be cool, but what is even cooler is the power source—or the *what*—that fuels it.

Consider the following:

You are up late online, typing, clicking, scrolling, and reading, but amidst such activity you are getting very little done since your mind is hazy and scattered; your body is tight and lethargic. There were times earlier in the day when you worked effectively, but now your attention is weak, worn-out from an active day of work or school. While it may appear to an outside observer that you are making progress on an important project, you are accomplishing very little.

A voice in your head beckons, "You have already been on the computer ten hours today. Why not do some stretching, read a book, or take a warm bath before bed?" You agree in principle, but you just can't pull your attention away from the social network, game, or news in front of you. Though you are choosing to use technology, you really feel as if it is controlling and using you.

The results of such long hours on the computer continue to impact you. Once you do finally disconnect, you are mentally exhausted and make your way to the refrigerator—the old, scattered mind state you are in believes that the right food will ease some of your tension and be a reward for your day of hard work. But in this frame of mind, you choose unhealthy food that is difficult for your body to digest, and you proceed to eat it with the same dazed attention. Trying to sleep some time later, your mind is still cluttered and your stomach bloated by your kitchen visit. You again turn on some technology—your TV, iPod, or computer—but nothing helps you relax. It takes you a few extra hours to get to sleep, and you wake up the next morning lethargic and drained.

According to one study, a typical information worker who sits at a computer all day turns to his e-mail program more than fifty times and uses instant messaging seventy-seven times per day.[6]

Though most of us need to be networked for much of the day, for those of us constantly connected we have many times like this example when our engagement is habitual, when it feels like technology is using us instead of us using it. While it is extraordinary that we can be connected, we often forget to know *what's* connected—or more precisely, what state of our mind is engaged. When this is forgotten or disregarded, our health and well-being pay the price. I learned this the hard way some years ago while sitting in the emergency room.

"Just how long do you spend on the computer each day?" the doctor asked me. It was eleven p.m. and I had come to the ER because, while working at my computer, my hands started twitching so much I could no longer type. The nerves in my body were shaking as if they were getting slight electrical shocks every few seconds.

The doctor had already worked his way through the first round of questions: Had I taken any illicit drugs? Did I eat anything out of the ordinary? Had I been working with chemicals? Was I on any medications? When I answered no to all of these, he followed with more general questions, like how I spend my days.

"I am on the computer about twelve hours a day," I finally responded, underestimating a few hours. The doctor's head tilted back slightly, his eyes widened, and he expressed concern. What century is he from? I thought. Don't most people work this much on

the computer? The doctor took my blood and ran some tests, all of which came back normal. By the time I left the hospital at roughly two a.m., he could find nothing wrong with me.

Sure enough, after taking a few days off, my symptoms went away. The cause of them, I surmised, had been my marathon computer use. My hours and hours online and linked in had caught up with me, such that my body was screaming to *link out*. Though I was connected and working very long days, I was paying almost no attention to my state of mind, and I was suffering the consequences.

I know I am not the only one who has felt the effect of such a relationship. In an article in the *New York Times*, blogger Michael Arrington, founder of the renowned blog Techcrunch, told of gaining thirty pounds and developing a severe sleep disorder in his constantly connected life, concluding, "At some point, I'll have a nervous breakdown and be admitted to the hospital, or something else will happen."[7] Sadly, this is likely true for more and more of us who put *that* we are connected and *that* we are working over *what's* connected and *what's* working. We increasingly feel the effects of an imbalanced and largely unconscious constantly connected life and are in danger of what my friend somewhat jokingly calls "death by computer."

The Internet is a particular challenge to find this balance with since it never closes and is thus always an option. We can now spend all of our waking life connected to technology. This is both great and terrible. It is great because no matter if we are a morning or evening person, the Internet is available. At any time we can check Yahoo! News, YouTube videos, or updates to a friend's profile page. However, this is terrible because there is really no time we can consider not going online. Many of us live with the ever-present voice, "You know, you could always go online. I wonder if someone sent you an e-mail or if so-and-so is online." We really *can* go online anytime. Before breakfast, "You could go online"; after breakfast, "You could go online"; before dinner, after dinner, pretty much any time of any day, we have this option. While the online world can help us find useful information to deal with many of life's challenges, it can also be an escape from, rather than an exploration of, our life. Our time networked, either by phone or computer, becomes compulsive rather than creative. We spend many late nights online but with little to show for it because our attention is only partially present.

When we focus on *that* we are connected, the technologies become more important than their function. We think, *Wow. Look how many videos I can watch. Look how much information I can consume in a day. Look how many people I can communicate with. Look how often I can read my e-mail and text messages.* This is impressive, on one level. On another level, however, it is meaningless since the impact of these actions depends on the *what* that is present during them. If, for example, addiction and anxiety are the motivators, if these are *what's* engaging in our watching, reading, and interacting, then we are simply cultivating these qualities in our own life and bringing them into the world. If, however, ease and care are present, then we are adding these to our life and to the world.

The needed shift is to focus less externally, less on the *that*, to the internal or the *what.* When we do this, our priority is no longer on, "Look how much I can consume and be connected," but is instead directed inward on, "What am I actually cultivating in this moment? What am I bringing into the world?" We see that no matter how sophisticated and awesome the gadgets in our lives, no matter how cool the *that*, our state of mind—or the *what*—always matters more. It determines the level of our ease or stress, and of our creativity or addiction. When we make this shift, our priority becomes the level of consciousness with which we are engaged.

We could see this consciousness as similar to the power source in any technology. The most advanced computer or sophisticated cell phone is fairly useless without a power source. Sure, we could use our non-charged computer as an end table or our cell phone as a paper weight, but they only come to life when they are plugged in or charged. The same could be said of our relationship to them. The fact that we can get online, download, listen, play, share, talk, send, and everything else is great, but like a technology without a power source, it is not all that impressive. What makes such actions really shine or not is the state of mind we bring, the presence in which we engage with them. When distraction, fatigue, frustration, or anxiety are engaging, we are like a cell phone without a charge. Our body's only use is in keeping our chair warm.

In this tool, when we are zoned out in front of the screen instead of continuing to work, paying almost no attention to the *what*, we bring consciousness to our experience. We acknowledge, "There is little focus in my effort at this time. I am primarily engaged compulsively instead of creatively." In this seemingly simple act, a world of possibilities opens. We have moved from unconsciousness to consciousness just in knowing what is engaging in that moment. From this, we can decide what action, if any, to take to increase the power source and to make our time more effective.

APPLY TO LIFE

The next time you are on the computer and notice that your attention is weak, try the following:

- First, inquire as to your state of mind. What is present? Is it excitement or boredom, connection or loneliness? It does not matter what you are doing on the computer or other technology—checking your e-mail, writing a letter, calling a friend, or surfing the Web—what matters is your state of mind. If you are not sure, tune in to your body. Are your shoulders lifted, belly tight, neck hunched over, and eyes glued to the screen? Or is there ease in your body? All these can be signs to help you know *what* is engaged at that moment. You do not need to judge yourself as good or bad, but instead simply bring awareness to what is true in that moment.

- If your actions feel compulsive or you have little focus, explore your options. Ask yourself, "What would be best for me in this moment? What could most serve the power source, the state of mind, I am bringing to this act?" See if it feels best to keep working or to shift your attention from the screen—take a walk, go outside for a few minutes, read a book, or write in a notebook—and refresh your body and mind through another activity.

Initially, this practice will seem difficult. However, every time we look inward to our level of presence, we are shifting from the habitual and unconscious to the creative and conscious.

TOOL #2: SEE CHOICE

You Mean, I Have a Choice?

It turns out that we are the masters of the technologies, not their servants. Who would have known?

Some time back I was enjoying dinner with an old friend at a little Chinese restaurant in San Francisco. I had not seen him for a few years, and we had a lot of catching up to do.

We were having a lively discussion when, about fifteen minutes into our meal, his cell phone rang and without hesitating, he answered it. He proceeded to carry on what seemed to me a pretty trivial phone chat. "Oh, I'm not doing much, just sitting here having dinner with a friend," he said, "What are you doing?" He went on to talk for several minutes while I sat there.

This would not have been so troublesome if it had only happened once, but throughout the night the same thing occurred many times. We would be talking, his cell would ring, and he would leave mid-sentence without a second's pause to answer his phone. It was as if once his cell rang, I disappeared completely. I kept thinking, Does he even know I'm here? Are all these calls so important that he can't let any go to voicemail?

I wondered if we would have a better conversation if I walked outside the restaurant and called him on his cell, since that seemed to warrant more of his attention. The fact that I was with him in person demoted me to a secondary priority.

According to researchers at Ball State University, "about 30 percent of all media time is spent exposed to more than one medium at a time."[8]

After this, I realized how often I did the same thing. Once I got a call on my cell, the people around me disappeared. My mental editing tool cropped them out. I vacated them for the person calling all because . . . the other person was trying to contact me through technology! It really made no sense. I then discovered the second tool in the Wisdom 2.0 life: *See choice.*

Similar to my experience at dinner, we often act habitually without any real thought as to where our attention is most needed at a given time. This is not bad or wrong, but it does not lead to what we seek: to consciously and creatively engage with technology. To achieve this, we need to see the importance of our conscious attention—what the Buddha called "our greatest treasure." In fact, where we direct it often expresses what we most value. If we answer our cell while out to dinner with our friend or partner, we are essentially communicating to him, "The person calling deserves my attention more than you do." Not answering it expresses, "You deserve my attention more than the caller." Of course, I don't know what in any given moment most needs your attention, but I do know that you can either see choice and be a master of these technologies, or not and be their slave. You can answer the phone habitually, thinking, *It's my cell phone for God's sake. I must answer it! I can't let it down.* Or you can see choice in that moment, realize that you are the master, and instead explore, *Where is my attention most needed at this moment?*

If we do not see choice and are a slave to these technologies, we have many situations like the following:

> *You are driving down the freeway at seventy miles per hour when your cell rings. I must get that, you think, though you are not expecting a particular call.*
>
> *Uncertain where your phone is, you start searching for it in the car, with one hand on the steering wheel as you only partially pay attention to the road.*
>
> *Where is that darn phone? you think as you reach under your seat, then stick your hand in your bag to see if you can feel it, trying*

to follow the sound. You glance in your briefcase but still cannot locate the phone.

You then look up and notice that you have swerved into the other lane. You quickly slam on your brakes to avoid hitting the car in front of you. A police officer notices this and pulls you over.

"You were paying almost no attention to the road," the police officer says. "What was going on?"

In such a moment, the most honest answer would be, "My master beckoned. I had no choice but to respond."

The master, of course, is not the caller, but the technology itself, the cell phone informing us, "You have a call." This master, in fact, can be so domineering that we cannot wait thirty seconds to pull over to the side of the road to find our cell safely and return the call. It demands our immediate attention.

I have never been stopped by a police officer or been in an accident while searching for my cell, but I have had several close calls, times when my attention was not focused on the road. If someone were to ask me, "Where is your attention most needed when driving down the highway at seventy miles per hour?" I would reply, "The road." I would answer that my safety and that of those around me is more important than taking a call that I can return thirty seconds later after pulling safely to the side of the road. However, if I do not see choice, then I have none. If technology is my master, I must heed its every call, and everything else is secondary.

In this tool, we practice seeing choice so we can direct our attention where it is most needed at a given time.

APPLY TO LIFE

The next time your cell phone rings while you are driving down the road or spending time with a friend, see choice by doing the following:

- Notice the sense of immediacy in your body and thoughts, *Who could it be? Who is trying to reach me? What news may they have?* Feel the sense of not knowing and wanting to know.
- Realize that you do not know. It could be your doctor calling to tell you that you have a terminal disease, a lawyer informing you that a great aunt has left you a million dollars in her will, or a salesperson trying to sell you insurance. More than likely, it is simply a friend wanting to tell you about her day. However, in the end it is all just information. Be aware of the sense of urgency, the lure of the unknown.
- Next, take a few breaths and know you have a choice in that moment. Ask yourself, "Where is my attention most needed at this time?" Let that be your guide.

Incoming Message . . .

It is easy to change modes of communication, from e-mail to IM to cell; the real challenge is making sure we bring our consciousness along.

An old Zen saying reads:

When sitting, just sit.
When standing, just stand.
Above all, don't wobble.

In our age, we might change this to:

When e-mailing, just e-mail.
When talking on your cell, just talk on your cell.
Above all, don't talk while e-mailing.

For most of us, the talking, working, or surfing online are not what is stressful; it's the time we spend wobbling. It's the multitasking and unconsciously switching back and forth between modes of communication. After we see choice, at times we do need to take a call on our cell or send an e-mail to a colleague. In fact, much of our day is often spent switching back and forth between different modes of communication: we talk to our colleague sitting next to us, then pick up our cell to take a call, then e-mail a client, then text message a friend, then watch a video online . . . then speak to our colleague again. We sometimes switch modes of communication every ten seconds. This can be exhausting and stressful when such transitions are done unconsciously and habitually. We can, however, learn to *consciously change channels* so instead of draining our energy by continuously multitasking, we move with ease from one mode to another. This is the third tool. To do this, we must better understand how our attention enters and leaves our environment.

"When people try to perform two or more related tasks either at the same time or alternating rapidly between them, errors go way up, and it takes far longer—often double the time or more—to get the jobs done than if they were done sequentially."[9]

—David E. Meyer, director of the Brain, Cognition, and Action Laboratory at the University of Michigan

For example, much information exchange today involves updating friends and family on one another's thoughts and activities. This effort has found its expression on many Web sites. One is Twitter, which for those who are not familiar with it, is an online social network to microblog—or post short entries in response to the question, What are you doing? Many people use it to give play-by-play reports of their lives. Common posts might read, "I am having dinner with a friend (5 minutes ago)" or "I am watching a movie with Tim (less than 30 seconds ago)." In this way, a person's friends can visit the site to learn what he is and has been doing.

However, when a person submits what he is doing—or tweets—when he is with other people at the time, he is not as present for them. Of course, his body is still in the room having dinner or watching a movie with friends, but in order to write and send the message, his attention momentarily leaves. He has to change channels. This is nothing against microblogging and sites like Twitter (I also microblog). This is true anytime we are with someone and reach for our phone to make a call or send a text message. In all these activities we must, at least partially, disconnect from our immediate environment. (Of course, we also do this any time we daydream or pick up a magazine to read while we are with others, but when we do so through a computer or handheld device, there is a deeper level of absorption such that our attention is even less present in our immediate environment.) Such disconnection from our surroundings is not bad or wrong—in fact, it is often fun, necessary, and extremely useful—but to do so creatively and effectively, we must perform such actions consciously, with ease instead of stress, with patience instead of haste.

This other choice we have, reflected in the following Zen story, is one of prioritizing consciousness as we shift from one activity to another:

> *Some years ago, a Zen student was sitting with his teacher when the teacher received a much anticipated letter from an old friend. Knowing his teacher had been waiting a long time for the letter and must have been very eager to read its contents, the student started to excuse himself.*
>
> *"Stop," the teacher boomed. "Please stay. I will open the letter later."*
>
> *"Don't you want to open the letter now?" the student inquired, surprised. "I know how long you've waited for this letter, and news from your friend is right there in the envelope."*
>
> *"Yes," said the teacher, "I have waited long, but I cannot open the letter until I have conquered the haste I feel. Once that has run its course, I will open it."*

Driven by haste, we unconsciously change channels, believing that any new communication via e-mail or cell should override what we are doing in that moment. We take a call on our cell, then type an e-mail, and then focus on a report, yet we move so quickly and with such haste that we lose our attention in the process. Our mind usually races several steps behind our body, trying to catch up.

When we consciously change channels, on the other hand, we skillfully exit one mode of communication so we can fully enter another. Instead of focusing on how much we can juggle at one time, we prioritize our consciousness as we shift from one action to another.

APPLY TO LIFE

The next time you are spending time with someone and you get a call on your phone or need to e-mail or text message someone, if this is where your attention is most needed, try the following:

- First, do what you need to in order to exit one mode. Excuse yourself if you are with others, or come to a stopping place with your current project. Know that your attention is vacating one channel. Acknowledge this and take appropriate action to leave that channel.
- Next, bring your full attention to the new mode of communication. Be as present as possible for it, as if it deserves your highest and most complete attention.
- Notice the quality of your communication both with those physically present and with the person you are engaging with through technology. Is it deepened and more harmonious by this practice?

TOOL #4: SEE CONNECTION AS PERCEPTION

The Ultimate Wi-Fi

To best use the technologies of our age, we must expand our view of what it means to be connected.

In my effort to blend the path of wisdom and the path of technology, I often wondered why it was so hard at times to be away from my cell and computer. I asked myself why, when I forgot my cell upon leaving my house, did it feel like an emergency, and did I find myself thinking, *God, I need my phone. How can I be out and not reachable? What will I do now?*

In most of these situations, there was little real difficulty as a result of forgetting my cell for the time I was out. The calls I received in that time could easily be returned later, and if needed I could always (God forbid) use a pay phone to call someone. What interested me was the feeling of lack and incompleteness when I was not on my computer or with my cell. To better understand this, I realized that I first had to pay more attention to my experience when I was using them.

I discovered that when I was online, even if I was not logged in to a social network or chatting with someone, I felt connected to other people. There was a sense of being part of something bigger than me, of not being alone. The fact that I was online gave me a sense of belonging. It was the same with my cell phone. Even if I was not using it, by having it on me and knowing I was reachable, I felt linked to other people. Essentially, I viewed technology as what connected me to others and made me feel less alone.

"A human being is a part of a whole, called by us the 'Universe,' a part limited in time and space. He experiences himself, his thoughts and feelings, as something separated from the rest. . . . Our task must be to free ourselves from this prison by widening our circles of compassion to embrace all living creatures and the whole of nature in its beauty."

—Albert Einstein

This, I learned, was actually the root of my difficulty and addiction. If I viewed technology as what connected me, then it was only logical that when I was not engaged with it, I would feel disconnected. In seeing this, I better understood this feeling of lack. I realized it was no wonder I clung to technology so tightly, no wonder it was so hard to turn off my computer, no wonder I felt so lonely when away from my gadgets. I mean, who wants to be away from their source of connection? This is why when people suggest we spend less time online or on our cells, we often think, *What! And be disconnected? No way.* We want connection, not disconnection.

These technologies, of course, provide many useful functions in our life, allowing us to conduct business, read news, communicate with friends, and express our opinions with people around the globe. On a function level, they are fantastic. However, at a deeper level, as incredible as they are, they are also empty. *Empty of what?* Empty of inherent meaning, of providing us true connection or satisfaction. This is no problem of technology; it is not that Facebook, MySpace, and Apple need to create better ways for us to connect—they are doing a fine job at what they do. It is instead because nothing external can ever satisfy this need. This is simply how the universal social network of life was set up. These devices are extraordinary when we look to them to provide various functions, and utterly disappointing when we seek true connection in them. This requires something else.

For example, we can enter a grocery store and feel connected to the cashier, or not; we can walk through an apple orchard and feel connected to the trees and birds, or not; we can participate in an online social network and feel connected to other users, or not. We can never touch a cell phone or go online and feel connected to ourselves and the world, or spend ten hours a day online and not feel this connection. Technology is not what determines this; how we view and relate to the world does. This is the fourth tool: *See connection as perception.*

The great myth, of course, is that our inner world does not matter and that technology, particularly the online world, can satisfy this need through providing news, social networks, and entertainment. I mean, what else do we need, right? I'm sorry to be the one to break the myth, but there is more to life. At some level you have always known this. You too have felt this same lack, this same disconnection, even after you spent hours and hours "connected" online. This experience can be confusing since it is so rarely acknowledged in our culture. Unlike cigarettes, our devices and social networks do not come with warnings on them such as: "Beware. Use this only for its function. It cannot provide true connection. We are not responsible for any feelings of lack you may experience." It would be nice if they did, but I don't see this happening.

When we do experience this lack, we often think, *Since I felt connected while online and now I do not, the answer must be to always be online and connected. This way, I will never feel lonely.* We were often lonely while online as well; we just did not notice it. In an odd way, this desire to be connected all the time is right on the mark. Of course we want to be connected! However, though we are right that the answer is to always be connected, we are wrong to believe that technology can provide it. The answer, instead, is to expand our view of what it means to be connected, to see that it comes from us.

When we fail to see this, everything else in our life is set up in contrast: playing with our child, time with our parents, a hike out in nature, sex with our partner . . . or doing just about anything else will always take a backseat to the gadgets in our lives. This view is what so often prevents us from seeing choice and is at the root of our habitual and addictive relationship. It is why when walking in a park with a friend we will always answer our cell when it rings, and why four friends can be at a café together yet all of them are talking on cell phones or IM-ing on their computers. Connection through technology trumps the connection possible with the people in our physical environment because of our view that the former provides true connection. Ironically, though we are constantly connected to technology, we are also constantly missing true connection because we are looking for it in devices, in the external, instead of in our perception, or the internal. If we do not see this, we spend much of our day in a state that I call disconnectedly connected; we are connected to technology but largely disconnected from ourselves.

However, before you toss your computer over the balcony or throw your cell in the river and head for the woods to fast and meditate in order to find true connection, know that there is another option. When we see that nothing external can satisfy this need, we can then express this connection in all aspects of our life, including via technology. *What does this mean?* One simple exercise to illustrate this is to make your hand into a fist. Please do this now, focusing very little attention on the action. Now, make your hand into a fist again, but this time direct your attention to your hand and the process of closing it into a fist. Feel the energy in your fingers, notice the shift as they fold, feel the pressure as they do. Bring consciousness to this act.

The difference in this simple experience was the amount of consciousness you applied to the act. Though this may seem like a very small difference, it significantly changes our relationship to any experience. We can notice this when someone touches us too. When a person puts her hand on our back with unconsciousness and frustration, we can feel it, even if no words are spoken. Our body usually reacts by tightening and trying to get away. On the other hand, when she puts her hand on our back consciously and with care, we also feel it. Our body usually responds by softening and melting into the touch. The placement of the hand on our back may be the same, but the level of consciousness coming through the touch is quite different.

Nothing is particularly connecting or significant about making our hand into a fist or someone putting a hand on our back. In the same way, nothing is truly connecting about sending an e-mail or talking on our cell. However, we can infuse these acts with true connection through our consciousness. We could say that it is not *in* but *through* technology (or anything else) that true connection is found.

It does not come from any device or gadget, but since connection is everywhere we are present, this of course includes such devices.

In this tool, we expand our sense of what it means to be connected so it is not limited to technology. We can then be connected when home alone with no one to talk to, by ourselves in the woods, talking to the teller at the bank, and chatting online with our friends. Only then have we discovered true connection, what we may call "the ultimate Wi-Fi."

APPLY TO LIFE

The next time you are out without your cell (or if this never happens, try leaving it at home sometime), open yourself to a bigger sense of connection.

When you engage with someone you do not know, such as the cashier at a grocery store, be present with that person. Open your mind to the possibility that something connects the two of you. In doing this, you don't have to gaze deeply into the person's eyes or invite him out for tea or to be your friend on a social network. No need to tell him, "I am connecting to you." He may call Security.

You are also not *trying* to create a connection. It already exists. You are merely opening your mind to see it.

See if, by doing this, a sense of connection overflows to other areas of your life.

Daily Practice: The Breath

Though it can help to have approaches like those in the Apply to Life sections that we use in particular situations, we can also incorporate daily practices into our lives to help us decrease stress and increase our access to the creative mind. Throughout this book, I include a number of them. The first one focuses on the breath.

Almost every health program (and spiritual tradition, for that matter) encourages taking full, conscious breaths. Similar to drinking clean, fresh water, it is hard to go wrong and the benefits are numerous. The breath is a good focus of attention, as it can be a stellar indicator of our mind/body state. For example, think of times you've felt frustrated or angry. How was your breathing in those moments? Most likely it was tight and shallow. You felt your breath in your chest, and your inhales and exhales were fast. Next, reflect on moments of deep relaxation—after a delicious meal, receiving a generous gift, during an engaging conversation, or in connected lovemaking. How was your breath? Most likely, it was full and at ease. You felt your breath in your belly, and each full inhale was effortlessly followed by a full exhale.

However, we do not have to wait for such activities; we can breathe consciously and deeply no matter what the situation is. Try this right now. Sit up straight, and soften any tension in your shoulders and jaw. Allow the breath to be full and natural. Bring your attention to the breath for three full inhalations and exhalations. Notice your breath rise and fall once. Notice your breath rise and fall a second time. And notice the breath do so a third time. Notice the difference in how you feel from doing this. Of course, you are always breathing; in this short exercise, you are simply bringing awareness from the usual chatter of the mind to the act of breathing.

In one study, students who practiced breathing exercises reported an increase in concentration during exams, along with a decreased sense of test anxiety, nervousness, and self-doubt.[10]

Ultimately, how many breaths we take does not matter, but in this daily practice we begin with three full, conscious breaths at various times in our day: while waiting for our computer to boot in the morning, standing in line at the store, waiting for the bus or train, before eating, when going to bed, when waking in the morning . . . all are opportunities to breathe consciously.

In reading this, you may be thinking, *I am so busy; I don't have time for this—like I have time to breathe!* It actually takes no extra time: you are already waiting for something—and you are already breathing! It does not take more time; it is simply a matter of choosing to use that time consciously. To live, we need to breathe; the question is, how conscious and full will our breaths be?

While some effort is involved in this practice, if you find yourself trying hard to breathe, relax. Instead of forcing the breath to be full, simply encourage it to return to its natural state. Invite the breath to be full without strain or stress. Like most things, the breath responds to invitation much better than to orders. In one way, at such a time we could say we are "taking" a breath, but another way of viewing it is that we are simply bringing our attention to the natural on-going breathing process with an invitation for the breath to be full. It is more a shift of attention than a new action.

In this practice, feel the breath come and go, as if life is breathing you. Feel the breath move through your entire body. As the philosopher Chuang Tzu suggested, "True men and women breathe from their heels." Gentle, easy, full-body breaths.

If you want to expand this practice, spend five minutes every morning or evening (or both) breathing consciously. Below are some steps to help you do so:

1. Find a quiet place at your home or work (bathrooms can be great places in the latter) or sit outside under a tree. Do your best to limit noise and other sensory input as much as possible.

2. Sit upright on a chair, cushion, the ground (or toilet seat!). You want a posture that is both alert and relaxed, with your back straight but not rigid. You can close your eyes if you want or focus your gaze at a place on the floor or wall.
3. Take a few full breaths and feel the air move through your body. Notice how the belly rises and falls, the chest expands and contracts, and the air comes in and out of the nostrils. Be aware of the pattern of the breath in the body, without needing to force or change anything.
4. As you pay attention to the breath, be aware. Pay attention to whatever arises in your consciousness, be it a thought, sound, smell, or sensation in your body. If you notice yourself getting lost in the content of thoughts, daydreaming about your upcoming meeting with your boss, or worrying about a sales call later in the day, then allow the breath to again be the focus of your attention.
5. In taking your attention from thinking to the breath, the mind may counter, "What are you doing? This makes no sense. Give me attention." This is fine. There is no need to take a position for or against such thoughts. Just notice them and, for this exercise, allow the breath to be foreground.
6. Keep an attitude of openness and gentleness, of allowing whatever arises, while using the breath when needed to anchor your

attention in the present moment. Instead of focusing on the right or wrong way to do this, explore and make it a learning experiment.

7. When the five minutes are up, open your eyes or expand your gaze from your spot of focus. Look around the room (or bathroom stall) from a place of presence, taking in the sights and sounds with a greater alertness.

Try this every day for a week and see what, if any, impact it has. You can always increase or decrease the time as needed.

PART 2

THE HELP CENTER: REDUCE STRESS

When we are burdened by fears or overcome by frustration, we make very little progress on any effort. It does not matter how smart we are, how many degrees we have, or how many successful Web sites we have launched. With a body in knots and a mind in turmoil, we make a third of the progress we otherwise would with a more balanced and relaxed state of mind. In such moments, we often believe that if we just try harder or stare at the screen longer or purchase some new software, a breakthrough will come. The answer, we believe, is to push harder in our current state of mind, as turbulent as it is. However, a real breakthrough usually occurs not by continuing in such a state, but by changing the state.

In the Wisdom 2.0 life, our help center is not focused on how to use a program or software, but on ways to reduce our anxiety and stress so we can approach our efforts with more focus and ease.

TOOL #5: MAKE SPACE

Stress Well

It is not what arises in our mind, but the space around it.

How many times has a well-meaning person approached you when you were upset and said, "Don't be angry," or, "Just calm down"? If you are angry at the time, it can make you more so, and if you are not, it may just ignite your anger. The statement implies that certain emotions are bad or wrong. Instead of thinking we should never experience qualities like anger, the best strategy may be learning *how* to relate to whatever thought or emotion arises. This requires that we *make space* for our experience.

For example, if you put a couple drops of blue dye in a cup of water, what happens? The water turns blue. However, if you drop that same blue dye in the ocean, what happens? Not much. It has almost no visual impact. The same dye is put in, but the difference is in the amount of space or volume. The volume of the ocean is so great that the impact is almost unnoticeable.

The same is true of our minds. We can either have a little or a lot of space for our experience. If anger arises in us for whatever reason, one response is to fight our experience, to repress it and pretend it is not there. We may say, "Everything is fine," but we do not feel fine. We are upset that our computer keeps freezing, our connection to the Internet is slow, or that we cannot find an important document on our hard drive. That is the truth, but instead of acknowledging it, we suppress it; we keep it hidden. I am not referring to situations when we must decide whether or not to share our anger with another person. Some situations may require that we keep our anger hidden. The central issue in terms of stress is whether we acknowledge it to *ourselves*, whether we have internal room for the experience. This is primary.

At times, we do not accept what is true because the quality counters our identity, our ideas of who we think we are. We believe, *I am not someone who is angry; therefore I will not allow this. This is not me.* When the opposing quality to our identity arises, we think, *I am* this; *therefore I cannot experience* that. It is not the roles we play in life as a manager, father, or Web designer that create this, but the belief that these roles are who we are and should therefore determine what we experience. We tell ourselves, "I have all this money. I should not feel unsatisfied"; "I have this powerful position. I should not feel powerless"; "I am constantly connected. I should not feel disconnected"; or "I have all these friends online. I should not feel lonely." The truth, however, is what it is, no matter the reasons we think it should be otherwise. When we fight our experience, when we think it should be other than what it is, we have very little space or, referring to my earlier example, we have what we can call cup mind.

In one study, 80 percent of workers feel stress on the job, and 42 percent say their coworkers need help dealing with stress.[11]

I know this state well. I had long practiced meditation and taught stress-reduction programs. I then found myself living a very stressful life completely immersed in technology. Though I could live fairly stress-free under certain conditions, when it came to being networked through technology, the habitual and addictive part of me emerged. As this happened, my initial response was, "This can't be. I'm not that. I teach stress reduction. I am not stressed." However, the stress was simply the truth, no matter how much I thought it should be otherwise. Though I denied it, it still showed in my health, sleep, and relationships. In fact, the more I fought it, the more negative impact it had. I suffered from a bad case of cup mind.

Ocean mind is just the opposite. In it, we notice a quality like anger and allow it to be. We do not judge ourselves for experiencing it, nor do we try to justify it—reasoning why we have every right to be angry because of what someone did and how anyone in her right mind would feel the same way. Instead of following this train of thought, we simply focus on what is true: anger is present. In doing this, we allow more room. Like the dye in a cup or ocean, a quality like anger may arise in two people, but the space around it and the effect on each can be quite different.

One definition of stress is our fighting or non-accepting what is true in a given moment. Whatever is happening, we think, should not be. One definition of stress relief, then, is accepting and allowing our experience, no matter what it is. Any time we have an awareness and acceptance of the moment, stress decreases. Such acceptance does not mean we like or approve of what is occurring in any given moment. Instead, it means we shift from thinking, *This should not be happening. This moment should be other than how it is,* to, *This is what's happening. This is the nature of this moment.* In doing so, the moment may still be challenging, but the tension around it subsides. In this sense, fear and anger are not inherently stressful, but the stress is instead determined by how we relate to these qualities, the amount of space we give them.

When we accept instead of resist our experience, these potentially stressful qualities have less power. This can often be seen with the quality of fear. Some people think that to be confident or fearless means that fear never enters the mind. They then attempt to erect a firewall to prevent its appearance, thinking, *Fear will not arise. I will make sure of it.* However, just the belief that such qualities should not arise gives them power; it's what creates the lack of space, or cup mind. In fact, by thinking fearlessness is the absence of fear, we become afraid of the mind state, fear. As such, it's not fear but the *fear of fear* that is often the most troublesome and stressful. When it does arise, as it will, there is tension because a state of mind is present that we think should not be.

For example, if you have an important presentation to give to a group, one way to approach it is by trying to look confident and prevent fear from arising. You walk into the room thinking, *OK, be confident. Impress them. Show that you know what the hell you are doing. Don't look afraid.* In this approach, if fear does arise while you give the presentation, you will try to repress it. You will think, *Shit. No, this can't be happening. I am a confident speaker, not a fearful one. How did this fear get through the firewall I set up? Damn, better try to look confident anyway.* Not only do you still need to give your presentation, but now you have the extra task of fighting fear.

Another approach is to walk into the room and be open to whatever arises. In this, you focus less on the content of the mind and more on the space in which thoughts and emotions are received. If fear arises, instead of reacting to it by thinking, *Oh shit, fear. Now what do I do? How did this happen?* you respond, *Ah, fear.* You may not be happy about its presence, but you know it is not an emergency.

You see it as the truth of the moment, and through such allowing, you add more space—or, more accurately, the natural spaciousness of the mind is now unobstructed. Since there is less resistance and friction around it, the fear has less impact, making you more likely to give your presentation well. Oddly, by not fearing fear, you have more confidence.

In this tool, we practice giving room to, instead of fighting, our experience.

APPLY TO LIFE

The next time you have an event in which fear may arise, such as giving a presentation or a talk before a group, try this:

- Do whatever preparations you need before the event. When you enter it, focus less on what quality may or may not arise, and more on meeting with acceptance whatever comes up. If fear appears, acknowledge it, silently noting what is true: fear is present. Watch the impulse to personalize it by thinking, *How could I be afraid? How could this be happening to me?*
- Take a couple deep breaths and imagine the fear held in an ocean mind. Allow it to be. See it like a small raft on a vast ocean. Know that it can exist and you can still perform well.

A Dalai Lama Teaching

Qualities like anger will arise; the real issue is how much we continue them.

Another way we experience cup mind is when we encourage and become consumed by potentially stressful qualities such as anger and frustration. While one tendency is to push them away and suppress them, another is to identify with them, to let them take over and direct our actions. When this happens with anger, for example, it quickly spreads through our life in the form of insensitive comments, unhelpful criticism, and continuous complaining. Like a computer virus, the more we give it to others, the more others pass it on, and the more damage it does.

I experienced this some time back on a social network that was created by a foundation to help "make good things happen" in the world. Most users of the site were very well meaning and worked for nonprofits or were involved in social entrepreneurial efforts. People were generally supportive, providing one another feedback and encouragement on their projects.

After participating on the network for about a year, a number of us encouraged the founder, a leading Internet entrepreneur and philanthropist, to let the online community have a role in deciding how some of the foundation's money was awarded each year. Soon after, it was announced that as an experiment he would give twenty-five thousand dollars to whatever group or person the community chose. There were no strings attached and no process to follow. The online community, which consisted of several hundred active users and anyone could join at any time, was encouraged to work out the details. It was completely up to us.

This news was initially met with great enthusiasm (wow, someone trusts us to give away twenty-five thousand dollars), and everyone knew more money was possible if the experiment went well. Members began suggesting possible recipients. However, excitement soon turned to anger as people realized that their great ideas were not shared by others. The community was filled with comments like, "How can you not care about kids in Africa? They need the money the most. You are so insensitive to those in poverty." Or, "The world is on the brink of destruction due to global warming. That is what needs to be addressed most. I want to help all kids, not just those in Africa. You are so narrow-minded." It got personal and the virus of anger spread, infiltrating much of the community. It took many months and thousands of posts for the community to come up with unified suggestions for how the money should be spent. Worse than the lost time, though, was the conflict's impact on the community. It was amazing to witness how quickly a community of do-gooders became hostile and unfriendly, and I had fallen into the pattern as much as the next person. There were many lessons to this experiment, but the one I took away was the incredible damage unchecked anger and resentment can cause.

Researchers have found that severe and long-lasting stress can speed up the aging process by affecting the body at the cellular level.[12]

You may not have experienced this in a social network, but instead in a business, friendship, or partnership. At some point, anger took over and communication fell apart. Though we may notice anger more on the large scale like when it flares up in a group or community, if we pay attention, we'll see that it is hard for most of us to go a single day without getting angry at least once. Sometimes, the cause is not one big event, but a series of little ones throughout the day: the bus or train is late, a colleague fails to perform a task, an associate sends us a rude e-mail, our lunch shows up cold. Countless events day after day create frustration, and if we are not aware, these moments can build and overflow into the rest of our life.

So, we may ask, *how do we counter the momentum of qualities like anger?* A similar question was posed to the Dalai Lama at an event I organized some years ago. The gathering brought the Dalai Lama into discussion with formerly incarcerated teens and adults. In the dialogue, a participant described a situation in which he was treated unjustly, and he queried the Dalai Lama on the most effective way to respond. The Dalai Lama replied that in such a situation, it is very understandable that one would feel angry. The challenge, he said, is to keep from being "changed by the anger" or "propelled by it" and to ensure that "the continuum of that anger does not last very long."

When we are not aware, anger can easily propel us in ways we later regret. We support, in the Dalai Lama's words, "the continuum of that anger." We are motivated and affected by it. We show up at work angry because of traffic, so we respond to a colleague curtly, who then responds back to us the same, and we later carry this stress home and speak to our child or partner disrespectfully. We then find ourselves lying in bed at the end of the day wondering why we cannot relax enough to sleep. The anger that started in the morning is still present.

So, you may wonder, *What advice did the Dalai Lama suggest to the man at the conference*? He said that to limit the continuum of anger, it helps to rely on qualities to counteract it. He encouraged the man to cultivate qualities such as tolerance and compassion so that when strong emotions like anger arise, they do not compel him to act unskillfully. Anger may still be present, but other qualities such as patience and compassion can also be there to counter it. By doing so, the motivating force that propels us is not the difficulty of the situation, but the positive qualities present.

You may be thinking, *This sounds good, but after a colleague unjustly criticizes me, I am pissed and need to respond.*

Sure, but there are a number of ways to do so. One is to think we are the anger and to let it direct our actions. In this approach, we justify whatever we do, thinking, *I had to act that way. I was angry. I had no choice.* Since anger arose in us, we presume we are excused from all responsibility and have a free pass to act as we wish. We view our state of mind, in this case anger, as having complete control over our actions.

Another approach is to be aware of the anger, but to not see it as who we are. The fact that we can be aware of it proves there is some part of us that is not the anger. In this, we focus less outwardly on, "Who did this to me and how do I get them back?" and instead inwardly by inquiring, "What am I experiencing and how do I best respond to the situation at hand?" Just this awareness of what is true in that moment helps us to reduce the anger's momentum. By doing so, a curious shift happens: though the anger may still be present, by looking inward and exploring it, we are cultivating patience and acceptance in that moment. Like the drops of dye in the ocean mentioned in the previous chapter, the anger is held in a bigger structure created by our courage to see it clearly. By doing so, we are able to *counter with the positive.*

We could also see this as the difference between responding and reacting to situations. In reacting, the same energy we receive, we give back. We react to anger directed to us with equal (or often more) of it, and we feel justified in doing so. I mean, someone else started it, right? By doing this, however, we are not only inviting the other person to respond back in the same manner, but the stress from such a knee-jerk reaction will linger with us.

In responding, on the other hand, we see that anger may be present in our body and mind, but it need not inform and direct our actions. Instead of seeing the anger as who we are, we see that there is a part of us that is simply aware of the experience. This part is not angry; it's aware of the anger. This awareness enables us to respond instead of react to a situation. From this, a response can arise based on the needs of the situation rather than the intensity of the emotion we may be feeling.

We can then find that sweet spot that resides between suppression (*This anger should not be present; I am someone who does not experience such states.*) and reaction (*This anger should be present. It is who I am, and I am therefore justified to act in whatever way I want.*) Between these, the most effective response can be found.

APPLY TO LIFE

The next time anger arises—when your computer freezes for the third time in a day, your project stalls, or your manager pressures you on a project—try the following:

- Take a moment to feel the anger. What thoughts are on your mind? What does your body feel like? How is your breath? Notice the tendency to lash out at someone or something. Feel for a moment the physical and mental state you call anger. There is no need to judge whether it should or should not be present. It is present, and the first task is to be with it as it is.
- Explore what you might do to skillfully relate to the situation at hand, to not continue the momentum of anger. Know that you have a choice in how you act. Invite other qualities such as patience and compassion to be present, and inquire what it would mean to bring them forth in this moment. By doing so, you are not saying to your anger, "Go away!" Instead, you are simply inviting other qualities to be present as well.

Find that sweet spot between suppression and reaction. Try this and see if it impacts the influence and continuation of anger in your life.

Look Where the Light Is Not Good

That which is the most painful can at times be the least stressful.

In the shift to a creative instead of stressful relationship to technology, old patterns—ways of living that do not serve us—often need to be shed. We usually know what these are. There is no God of the Internet who has set forth Technology Commandments dictating just how long we should spend on our computer or cell each day. Though there are no external criteria, in each of us is an inner sense of what we need in order to live a more fulfilling and creative life.

We can make such changes in a number of ways. One way we often try is through judgment, by believing the thoughts, *You are bad for spending so much time online. You are wrong for using your Blackberry so much. You must go to bed sooner.* The quality behind these is criticism, and it often includes the words *must* and *should*. We feel we are bad or wrong for acting a particular way. When we believe these thoughts, we then seek to become good, by taking on a new schedule or health program that may work initially, and while doing it we think, *What a good person I am.* However, when we later quit and fall back into old patterns, we conclude, *What a bad person I am.* Of course, judgment can motivate, but it often creates more stress than it dissipates. In judgment, we are always on one side or the other, always good or bad.

In a study by the American Psychological Association, more than half (55 percent) of all adults in the study said they were less productive at work because of stress.[13]

Another way these changes can take place is through a deepening of consciousness, to *change through awareness*. The Buddha said that when we see that what we are holding on to is on fire, we will let it go. If we have a burning ember in our hand, for example, we could tell ourselves, "This is bad. You should not be holding this. Good people don't do such things." Other people could warn us, "You know, studies show that it hurts your hand to hold a burning ember too long," or, "My sister held a burning ember for a day, and she really did a lot of damage." Such information may be helpful to an extent, but no matter how much we criticize ourselves for holding the ember, no matter how much someone else tells us to let it go, until we fully feel the pain, we are not likely to drop it.

In the same way, when we feel the pain, isolation, and hurt in an action that is not aligned with our deepest well-being, we want to drop it. We want to stop downing bags of M&M's while working, we want to turn off our computer by ten o'clock each night, we want to exercise—not because we are bad if we do not, but because it feels right to our system. This is different than judging or berating ourselves. There may be a sense of "this does not feel right" but the change comes from a care for, not a hatred of, ourselves. In this way, unhealthy patterns drop in their own time as consciousness deepens.

This does not mean such change will always be pleasant. In fact, it can be very painful. If we have consistently followed a particular pattern, responding any differently can initially feel quite uncomfortable. That's the nature of change. It is similar to physical patterns. If, for example, every day the last year we drank five cups of coffee, and then one day we no longer do so, our body will likely experience significant pain in the short term, even though cutting back the caffeine may help us feel better in the long term. Letting go or no longer feeding destructive mental patterns works the same way. If the last one hundred times someone criticized us we responded angrily, to refrain from doing so the 101st time will likely be excruciating. It may take everything we have to keep our mouth shut and simply receive the person's feedback. At such times, we might think, *How can this be the right action? It's torturous!*

However, if we just follow what is easy or pleasant, we often miss the greater learning. There is an old story about a man named Nasruddin that illustrates this pattern. Nasruddin was a Sufi—a branch of Islam focused on inner knowledge and discovery. He is believed to have lived in the Middle Ages, though very little is known about him. Over the years numerous stories have been attributed to him. He is often in the role of a wise fool whose simple, innocent ways provide teachings for the rest of us.

> *One evening Nasruddin was looking through the grass in his backyard.*
>
> *A neighbor came by and asked, "What are you looking for?"*
>
> *"I am trying to find my keys," Nasruddin responded.*
>
> *"Oh, let me help," offered the neighbor, and the two men spent the next hour combing every inch of the backyard. Finally, the neighbor concluded, "We have looked in every possible spot back here. Are you sure you lost your keys in the backyard?"*
>
> *"Oh no," replied Nasruddin, "I am most certain that I dropped them while in my front yard."*
>
> *"Then why on earth did we just spend the last hour looking in your backyard?" replied the neighbor in exasperation.*

"Well, that is easy. You see, the light is better in the backyard so it is easier to look back here."

Reading this story, you likely think, *Jeez, what an idiot.* Sadly, though, we are often not much different. We know what helps us to live with less stress, but we often fail to act accordingly. A friend may say to us, "You know getting angry at other players in an online game only causes you more stress, so why are you doing it?" Or, "You know if you spend more than eight straight hours on the computer you get a headache, yet you still do it." Or, "You know that trying to send a text message while driving your car is dangerous, so why are you doing it?" Our response is often much like that of the man in the story. We are simply doing what is easy and familiar, not what is most effective. An honest response to our friend is, "I am doing this because it is familiar." Or in other words, "The light is better in the backyard." However, it only seems pleasant because we are not looking more deeply; we are not letting ourselves feel the impact.

When we let awareness guide us, we are less motivated by the thought, *This is not good. You should not do this.* Instead, we feel the "not goodness." Who knows? It could be that in this deepening of awareness, we discover that a particular pattern is not so painful, or that the pain we receive is worth the pleasure it also provides. Or it could be that through this deeper knowledge we feel the pain more completely, and a desire to suffer less arises in us such that the old pattern is no longer desirable. We simply do not want to hold the burning ember any longer. In this approach our entire system makes the shift, not just our mind.

We could say there are two kinds of pain. One is the pain of continuing an unskillful or stressful pattern; for example, of reacting by yelling back at the person who upset us. Another is the pain we experience when we no longer repeat a particular pattern. We feel this pain, for example, in not yelling back in anger after we have done so the last hundred times. The former we could say is the pain that leads to more pain, while the latter is the pain that leads to less.

In this tool, we look where the light is not good and feel the unpleasantness that can lead to more pleasant change in the end. We change through awareness.

APPLY TO LIFE

The next time you are in a pattern that does not feel healthy, try this:

- Notice any sense of frustration or self-hatred you may feel for having followed this pattern in the past. Forgive yourself. It is fine that you have continued this pattern as long as you have. There is no reason to hold any despair or self-hatred. Start fresh.

- Next, bring awareness to this state. If you find yourself habitually snacking on potato chips while you work and do not want to do so as much, bring awareness to the act. Often we experience a duality at such times, with one force saying, "I want to snack all day long. It feels so good." At the same time, a counterforce says, "You should not be snacking all day long. It's not good." We then snack, but feel guilty as we do, and by doing so not only do we enjoy the food less but we add more stress to our lives since we eat with guilt. One way to end the duality is to bring consciousness to the act, in this case snacking. We pay attention to the process with curiosity instead of judgment: noticing the desire for the snack, the reaching to get it, the putting it in our mouth, and the swallowing. We snack consciously.

As you bring consciousness to a pattern, see if another action wants to emerge, if it is time for it to change. It is fine if it does and fine if it does not. Notice the tendency to label it as good or bad and instead focus on your direct experience.

TOOL #8: JUST DO

You Must Be Logged in to Do That

Logged out of the moment?
No problem, just log back in.

I am often impressed by friends who hold numerous responsibilities—working a very active full-time job, volunteering significant time to a nonprofit, starting a Web company in their free time—yet when I spend time with them, they seem quite at ease. They are not busy or rushed. If I did not know the details of their life, I would never guess just how many responsibilities they have.

I have other friends, on the other hand, with as many or sometimes fewer responsibilities, but when I am with them our interaction feels heavy, they act as if there is something else they really should be doing, as if they are always behind on some important project. They continually look at their watch, check their text messages, fidget in their chair, and talk as if they never have any time—their to-do list is out of control and a great burden.

From this, I realized that our stress often has less to do with the length of our to-do list, and more with how we hold it, whether we carry or drop the weight of it.

Consider the following:

You wake up and before your foot touches the floor on the way to the bathroom, you are hit with the list of tasks you need to complete: the report to your manager that you must review and send off, the e-mail from a colleague that has been in your inbox for the last week and needs a reply, the call to the irate customer that you have been avoiding but know you can't much longer, and the social network you have not checked for days and has numerous messages awaiting your reply.

The list weighs on you as you attend to your morning activities. As a result, you shower quickly, down a strong cup of coffee, gobble up a pastry, and head out the door, hoping to start work early so you can get a jump start on this ever-increasing list.

I should have done these things last week, you think to yourself as you commute to work. How could I have fallen so behind? As such, you arrive at your destination agitated and scattered, beginning your short walk to your office.

If we are not aware, this pattern can stay with us throughout the day. *How*, we might ask, *do we attend to our to-do list but not carry the weight and stress of it?* To answer this, we need to know the teaching offered by the Zen master in the following story, a teaching we may call the art of just doing.

A renowned martial artist once went to visit a Zen master. The martial artist had spent years mastering his skills such that he was the toughest samurai in the land. He was an amazing swordsman and legendary for his ability to fight numerous attackers.

When he met the Zen master, the samurai talked about all the powers he had developed in his life, how he could defeat a hundred men in battle, jump on buildings, and perform other extraordinary feats. He then looked at the Zen master and said, "I have told you all the powers I have gained. You are well-known as a great Zen master, but what can you do? What powers do you possess?"

The Zen master took a deep breath, and then responded, "I only have one power: When I walk, I just walk. When I eat, I just eat. When I talk, I just talk."

You may think, *What kind of power is that? That doesn't sound so hard to me.*

To see how this may be useful, let's return to the previous example. When we left it, you were about to walk from the train station or parking lot to your office. There are a number of ways to conduct such a walk. One way is to carry the weight of your to-do list; the second is to drop it like the Zen master and *just do*. Let's look at how these two options may play out.

> **"It's no accident that things are more likely to go your way when you stop worrying about whether you're going to win or lose and focus your full attention on what is happening *right this moment.*"[14]**
>
> **—Phil Jackson, who has won nine NBA titles as a coach and is the NBA's career leader in playoff victories and playoff winning percentage**

In the first one, carrying the weight of the to-do list, you hurriedly walk to your office, thinking as you go, *Better get there fast. Then I've got to call that customer who is upset, can't forget that e-mail to Martha, and oh, that report, shit. I better think of what to put in it.* Filled with this mental activity, you miss the sights and sounds along the way, including the breeze on your face and eye contact with those you pass. You are weighed down by your to-do list so that even before you sit down at your desk and reach to click the power button on your computer, you are already stressed. You are logged out of the moment; you are missing it.

In the second option, dropping the weight, you acknowledge the list, write it down if that would help, and focus instead on just doing one thing at a time. In this case, on the walk to work, you just walk. You take your attention from the to-do list to the activity of the moment: walking. Of course, once you arrive at your desk, you know there are numerous phone calls to make, e-mails to send, and reports to review, but that is the future. It is not the activity of the moment.

In just walking, you feel your feet touch the ground, the air against your face. You notice the maple tree in the field and make eye contact with people you pass. Your breath deepens and your mind opens. Though you woke up in the morning weighted by the list, in just walking, you drop it. You log back in to the moment. The list has not been forgotten, but it is no longer foreground. Your walk is rejuvenating and refreshing. When you then sit down at your desk, you feel ease and vitality instead of tension. This occurs not because you completed any of the items on your list during the walk, but from letting the list go and just walking. In these two scenarios, the time and length of the walk do not change, but your experience does.

The impact does not stop here. Since the mind state in which we begin work is affected by the moments preceding it, how we walk also determines how we address our to-do list once at our desk. If we carried the list as we walked, then when we make that call to the irate customer, instead of addressing the needs present, we will likely be thinking during the call, *Do you know how many other tasks I have to complete? I am so behind. This needs to be quick. Hurry up.* Of course, we will not likely *say* this, but if we *think* it, the other person will feel it. He will feel the weight of our list hovering over the conversation. The customer, of course, does not want to be responsible for our list; he simply has an issue that needs to be addressed. If there is extra pressure in the discussion, the customer will feel he is not receiving the necessary attention. As such, he is more likely to react by complaining and criticizing.

If we drop the weight of the list and just walk, on the other hand, we are more likely to arrive at work refreshed and relaxed. We can address our to-do list by taking one task at a time. When we make that call to the irate customer, we can just respond. We know for that moment, that is what we are doing. Our attention is on the call, not on our list. With more of our attention available, we can better address the needs of the situation. By doing so, the people we encounter experience our attention and clarity instead of the weight of our list. As a result, we can more effectively complete our tasks.

We could see this as the difference between viewing a list and downloading it. Stress, for example, does not arise because we have the thought, *I have so much to do,* but from downloading that thought and identifying ourselves as someone with a lot to do. You can see these people in the hallways of businesses, rushing from one meeting to another, as if announcing as they walk, "I'm someone with a lot to do, so you must step aside. I have no time to wait." They believe this is who they are.

The other way is to simply acknowledge that there are many items on our to-do list. That may be what is true for us. Imagine talking to someone and when you ask how she is doing, she responds, "There are many items on my to-do list." She says it simply as a matter of fact, without a lot of tension around it. It is simply how life is in that moment. She says it from a place of ease. However, when you ask the same question to another person, he answers, "God. I just have so much to do," and he goes on to complain for fifteen minutes about how crazy his life is. In this person's response, there is heaviness and tension. The to-do list is downloaded into his system and carried wherever he goes. Instead of just doing, he is weighed down by the list.

The power of just doing may not seem like much, but we can compare it to the headlights on a car at night. We could think, *They only let you see fifteen yards in front of us. That is not a lot. How much help can they be?* However, with only that amount of visibility, we can travel across the entire country. All we really need to see is what's directly in front of us. This may also be true in our larger life: we can do more not by focusing on what is two hundred miles ahead, on what we need to accomplish later today or next week, but on what is in the next fifteen yards, on the task at hand.

In this tool, we don't drop our to-do list, but we drop the weight and stress of it. We accomplish a lot by focusing on the little in front of us.

APPLY TO LIFE

The next time you are walking to your office, whether it is across the living room or after getting off a train, just walk. Sure, you may have one hundred things to accomplish once you get to work, but in your walk to your office, enjoy walking. Make it a meditation.

As you do it, if the thought arises, *Remember to call Jim,* there is no need to fight this with, *I am not supposed to be thinking that. I am supposed to be just walking.* It is fine if the thought arises. Register it, and then bring your attention back to the walk. In fact, you may find that through just walking, you actually remember more items that need your attention. You can register one, then another, without getting lost in any, without logging out of the moment.

As you walk, feel the breeze against your face, hear the sounds of traffic, see the leaves of the trees rustling in the wind, and appreciate the grass or trees around you. Be present for the walk, knowing that the mind state in which you walk will affect how you begin work once you arrive at your desk.

The Life Alignment

Because the outer responds to the inner, sometimes the best way to address the external is to align the internal.

The times in my life when I most habitually and stressfully related to technology, I was also most out of touch with what mattered to me. If asked, I would answer that my health, eating well, and spending time in nature mattered, but my actions said otherwise. If you watched my actions, these were not at all my priorities. By my actions, what most mattered to me was knowing if I had received any messages in the two minutes since I had last checked. That is how I spent much of my day.

This disconnect, I realized, was adding difficulty to my life. I thus had to either change what mattered to me or change my actions. By feeling one thing internally while my actions expressed completely different priorities, my life was out of alignment. In this situation, the very things that were truly important to me, like my health and well-being, were not being addressed. I realized that no matter what I said, my actions had consequences.

Consider the person in the following story:

A young woman once walked into an Internet café and saw a man with gray hair and a wrinkled face sitting hunched over at a table, actively engaged with his computer. He was typing fast and moving his mouse around with great speed.

Wow, she thought, that old man seems quite skilled at computers. Sitting at a table behind him, she noticed he was playing World of Warcraft and doing very well. A few minutes later, the man took off his headset and went to pick up a cup of coffee. On his return, the woman said, "I was watching you, and I noticed you're quite skilled with computers. What's your secret?"

"Ah," said the man, as he slowly lowered himself into his chair, his bones creaking as he did, "I play twelve hours a day, almost never go outside, survive on Doritos and Twinkies, and down a cup of coffee every hour."

"Impressive," said the woman. "And you have managed to live to a ripe old age while being so tech savvy. If you don't mind my asking, just how old are you?"

"I'm twenty-eight," the man replied.

This story is kind of funny (at least I think so), but it does have a point: our actions have consequences. In telling it, my intention is not to lay judgment on the choices people make. The man in the story may be living a very conscious life or he may be living largely unconsciously. Either way, these are his choices. However, for those of us who seek to live consciously and with less stress, it helps to align the internal and external, to see if our external actions do justice to what internally matters to us.

In a study sponsored by Sheraton Hotels and Resorts of 6,500 working U.S. professionals, 35 percent said they would choose their PDA's (personal digital assistants) over their spouses.

For example, if someone says, "My health and well-being are important to me," and she lives like the man in the story, there is little alignment. And if she says, "Eating sweets and playing online games are all that matter to me," and lives that life, her actions have integrity. Other people may agree or disagree with her choices, but the actions are aligned with what matters to her.

So, what matters to us? The following story is one person's answer:

A techie guy was walking down the street in a rare break from his computer and noticed a frog on the ground.

"If you kiss me," the frog called out to him, "I will turn into a princess and stay with you forever."

The techie picked up the frog, smiled at it, and put it in his pocket.

A few minutes later, the frog spoke again. "Didn't you hear me? I said that if you just kiss me, I will be yours. In fact, I will be completely devoted to you and be the best sexual partner you can ever imagine. I will satisfy your every desire."

The techie looked down at the frog again and kept walking.

Finally, in desperation the frog asked, "What's the matter? I am offering you so much devotion, so much love, and so much sex. All you have to do is kiss me."

The techie responded, "Listen, I am a techie. I spend all my time on the computer. I don't have time for sex or a girlfriend, but having a talking frog . . . that's cool."

This, of course, is a joke, but it brings up the question, What really does matter to us? If we live like the man in the café, aging faster and surviving on Twinkies and Doritos, the central question is, Is this aligned with what matters to us? If it is, fine. If not, then some adjustment is needed. We can make such changes not because one way is bad and another way good, but because one way is more aligned with what truly matters to us.

As this adjustment takes place, as we better understand, clarify, and give attention to the internal, we often experience a shift in the external. How many times, for example, have you greatly desired something and not been able to get it, but the moment you let go of it, the moment you released the desire, it came to you? Through an internal shift, the external responded. Or after months looking for work and finding nothing, the day you became very clear on exactly what kind of job would be best for you, it appeared. Of course, even with such an alignment, struggle and challenges still arise, but we can handle them better since we have deeper integrity and stability to our life.

In this tool, when the outer is filled with chaos and stress, we look inward to see if what matters to us internally is aligned with our actions externally.

APPLY TO LIFE

Ask yourself, "If my inner and outer life were aligned, what would that look like? How might I live if this were true?"

Explore what it might be like to have such alignment. Know that doing so does not necessarily require huge changes to your life, but can also include the perspective and approach you are taking. See what needs greater clarity to help create such an alignment.

Daily Practice: Eat to Eat

I don't know the exact reasons, but in times of frustration and difficulty, we often think eating will help the situation. Can't figure out a coding issue on a Web site glitch . . . report not coming together . . . manager angry at us for making a mistake . . . if we just eat, the issue will be solved. In such challenging moments, we think about—and often consume—the cookies in our bag, a muffin at the snack bar down the street, or the candy in the machine at the end of the hall. Of course, eating a healthy meal may be beneficial in such moments, but when our eating arises from frustration or compulsion, it rarely helps—and often hinders—the situation. In fact, eating is probably one of the most difficult areas with which to bring consciousness.

In our constantly connected world, we often eat while performing another action: walking from one meeting to another, reading news online, working on a project for work, or carrying on several online chats. This may be needed at times, but by consistently doing so our life gets more and more rushed, and we become increasingly stressed. We are so busy that we cannot take time to enjoy one of the great pleasures of life: eating. When this happens, we eat our tension more than we eat our food.

The consequences of this are many: we do not chew our food so our digestion is not good, we eat more than we need and gain weight, and our energy is low since more work is needed for digestion. Because we often eat while doing something else, we also fail to taste the food. After a day of eating like this, we go to bed utterly exhausted.

Of course, *what* we eat impacts us. If we eat less sugar, processed foods, heavy dairy, and meat products, and we eat more fruits, vegetables, nuts, and whole grains, we will likely have more energy and mental clarity. Another factor, however, is *how* we eat. Do we gobble our food while sitting in front of our screen playing Half-Life or watching a YouTube video, rarely paying attention to what we are consuming? Or do we find a quiet place and enjoy each mouthful?

In a study of 1,848 adults conducted by the American Psychological Association, during one month nearly half of all respondents (43 percent) overate or ate unhealthy foods, and more than one-third (36 percent) skipped a meal because of stress.[16]

Most of us know quite well how to eat unconsciously: eat as fast as you can, doing as many other things as you can, in as noisy and hectic an environment as you can find. That's easy. In this daily practice, we do just the opposite: we eat slowly, don't do other things at that time, and eat in the most peaceful environment we can find.

Such conscious eating may be hard for three meals a day, so in this daily practice start by picking one meal—either breakfast, lunch, or dinner—to eat consciously. Initially, it may take all your effort to pry yourself away from your computer screen, but many people find that when they bring their attention to the process and taste their food, they enjoy food more than ever. While it can be difficult at first, it can also be much more pleasant and enjoyable and can help us return to our work with greater energy and focus.

If conscious eating feels like a chore, make it an experiment. One day, eat quickly and unconsciously, and on another eat slowly and consciously. Play around until your system finds what is best for it. Instead of thinking there are good and bad ways to eat, experiment and let your body guide you to what is best. In doing so, if you find yourself unconsciously downing M&M's in the middle of a workday while your eyes are glued to the screen, notice how judgment often arises. Instead of staying with the judgment, take a few breaths and sense what your body and mind need in that moment.

If we pay attention, our relationship to eating is often less about nourishing our body or how the food tastes and more about satisfying our desire system. You may notice that even before you have chewed and swallowed one mouthful, you already crave the next bite—even though the desired next amount is exactly the same as the one in your mouth. I notice this with my favorite snack, cashews. As soon as I put three of them in my mouth, I start craving the next three I see in the bowl, even though I have yet to enjoy the ones in my mouth. Oddly, I want what I already have! In such times, what we crave is not the food, but the satisfaction of getting what we want, of having our desires filled. Ironically, however, our desire is rarely filled because the moment of satisfaction is so fleeting. It exists only at the time when our desire for food is satiated by putting it in our mouth. That's it. Once it is there and before we have chewed and tasted it, the desire for more arises.

In conscious eating, we give attention to the full process of eating, receiving satisfaction not only in the moment we put the food in our mouth, but also in the chewing, the tasting, and the swallowing of it. We enjoy all stages of the process.

Below are some suggestions for your one conscious meal per day:

1. Find a quiet place to eat with as little stimuli as possible. It can often be nice to be near nature, if possible.
2. Clear the space in front of you so your attention is less likely to be drawn to work-related items or various media. Move away from your computer and put away magazines, cells, and other material.
3. Choose food that is healthy and good for you.

4. Before eating the food, open your sense doors. Notice the sight of the food, the colors, and texture. Notice any smells that arise. If you have finger food, like a sandwich, feel the texture of the bread. Let your other senses experience the food before tasting it.

5. Take three full breaths, stabilizing your mind and body.

6. Next, as consciously as you can, chew and taste the food. Notice the flavors. If it helps, put down your fork or the sandwich as you chew. Finish one bite before taking another.

7. Welcome the food as nourishment, as an energy source for your life.

THE PAUSE

To get somewhere more quickly, know how to stop.

Though it is hard to know just how much time we spend waiting, one report estimates that if we live to be seventy, the average person will spend three of those years waiting. Though that may seem like an overstatement, think of all the times we wait—in lines at stores, at red lights, for Web sites to open, and for videos to load. Whether the number is three years or one, even for those with exceptional time management, life is filled with waiting. We are continually forced to stop.

I don't know about you, but the more I am in a hurry, the more life makes me wait. The later I am when driving to an appointment, the more likely I will get stuck following a slow-moving car; the more I need a Web site to open quickly, the slower it does; and the more I try to hurry in and out of a store, the longer the line is. Don't ask me how this works, but the more I rush, the more life seems to make me stop and wait.

When we live in a rush, the tendency is to meet moments of stopping with the popular mantra: "Shit."

The stoplight turns red as we hurry home after a long day of work: "Shit."

We can't get online because our connection is down: "Shit."

We walk into Starbucks and see a fifteen-person line: "Shit."

Someone steps in front of us as we walk to the door, making us pause for a few seconds: "Shit."

We must stop working on a project due to an error by a colleague: "Shit."

A Web site is taking longer to load than expected: "Shit."

I have had many of these days, ones where I carried this "shit" response from one activity to another, creating a day full of "shit."

Though we tend to fight these moments, at the same time we often complain about how few breaks we have in a day, telling our friends, "My days are so busy. I have almost no free time to relax." Yet if we look more closely, we had various opportunities; life stopped us many times. We waited three minutes for an elevator, sat in our car for a minute at a red light, waited ten minutes for a late colleague, and looked at the screen for thirty seconds while a video loaded. All were potential breaks, but instead of reducing stress, they added to it as we spent that time thinking, *Where is that elevator?* or, *Yet another red light*, or, *When is my colleague finally going to come?* or, *Come on, load already!* The stress, of course, is not inherent in the stopping, but arises in our response to it.

The stress around stopping often comes from a discomfort with what we may call our inner life—how we experience ourselves internally. The more out of touch and uncomfortable we are with our inner life, the more difficult stopping becomes. For example, if we are late to a meeting with a person who has such a relationship, in those five minutes he's waiting, we are asking the person to do what he tries to avoid: being present with his own mind and body. When he has to stop, the disconnection from his inner life is more distinctly felt. He feels the hollowness that can usually be kept at bay through living a very busy life.

"It takes, on average, 16 min. 33 sec. for a worker interrupted by an e-mail to get back to what he or she was doing."[17]

In such situations, people often believe their discomfort is caused not by internal conditions, but external ones. Someone, they believe, is making them feel hollow and lonely. Thus, they get angry at us for showing up late, at the long line in the store, at the Web site loading the video, at whatever is forcing them to stop. However, the stopping does not create the tension; it reveals it. It uncovers what we can often avoid through constant busyness. This is often the real source of our frustration. So, when we think, *Shit, he's ten minutes late,* or, *That damn line,* the deeper message is, *Why must I feel this discomfort? Take it away, please.* But in looking externally, we never touch the root of our stress.

Of course, there may be issues, like a colleague's continual tardiness to meetings, which need to be addressed. However, if we do not blame the person for our discomfort, if we do not bring our dissatisfaction with our inner life into the issue, then we can more skillfully address the problem. We see that the situation provides an opportunity for learning, and we can address it without the extra frustration.

This does not mean that we can use this as an excuse for our own actions. The next time we are criticized by our manager for showing up late to a meeting, it is probably not best to reply, "Well, if you weren't so uncomfortable with your inner life, this would not be such a problem. Deal with yourself." I don't know your manager, but

I don't think that would go over well. Of course, you can do as you wish, but we are only responsible for our actions, not others'.

In this tool, we do not look to spend time in lines or visit slow Web sites, but when we must stop, we see that we have a choice: to meet it with "shit" and look for someone to blame, or to accept it with "ah" and use that time to take a break. In the latter, instead of days full of shit, we have days full of *ah*s. We no longer view stops as mistakes of the universe, but as invitations to breathe deeply, to relax our body, and to open our senses. Through this tool, those three years of our life we spend waiting will more likely be times of ease instead of stress.

APPLY TO LIFE

The next time you must stop, be it while stuck in traffic or when your computer crashes, welcome it. Meet it with, "Ah."

I can hear you thinking, *Yeah, but when I am put on hold on the phone for a long time or walk into Starbucks at ten a.m. seeking my morning cappuccino and see a twenty-person line, I don't feel* ah. *I feel* shit.

OK, point taken. Me too. So I started a practice for us. It is called the Ah, Shit Practice.

When you must stop, start by adding "ah" to "shit." At first it's "ah, shit," but then extend the *ah* each time, such as . . .

"Ahhhhh . . . shit."

"Ahhhhhhhh . . . shit"

"Ahhhhhhhhhhhhhhh . . . shit."

"Ahhhhhhhhhhhhhhhhhhhhhhhhhhhhh . . . shit."

Gradually, we have more and more *ah*s, and less and less *shit*.

During the time you are waiting, explore viewing it as if the universe in its great compassion is giving you a much-needed break. Instead of focusing on all the ways the moment is not right and getting caught in frustration, take three full, conscious breaths and relax any tension in your body. Then open to the world around you. Notice the other people around you and the setting, including the colors, shapes, and sounds. Explore what it might be like to wake up and be present as you wait. Notice the difference in how you experience your time waiting as you do this.

TOOL #11: DRAW A LONGER LINE

The Only User We Can Control

In any business, game, or social network,
there is only one user we can control.

We all have certain people in our lives. You know the ones: the people who, for whatever reason, you detest and do everything you can to avoid. Yeah, those people. In our technologically rich culture, they show up in more places than ever: at work, on social networks, in online games, in e-mails, and within comment sections. Sometimes, how certain people act is what irritates us: customers who want special treatment and send us wordy, incoherent e-mails; managers who never have a word of praise no matter how well we do; online users who are belligerent and insensitive; or colleagues who, time after time, fail to follow through with their commitments.

Other times, their personality or mannerisms are what irk us: the arrogant or haughty way they walk into a room or the lisp in their voice that makes our body react as if someone has just run fingernails down a chalkboard. Sometimes, in simply looking at them, we cringe. Or if we interact with these people online, the way they use words or emoticons irritates us. Such people may be difficult personalities who our friends and colleagues also find annoying, or they could be people whose systems just are not compatible with ours. Since most of us must relate to people not of our choosing, we all likely have such people in our lives, be it a sibling, professor, boss, employee, customer, or online avatar. It is easy to view such people and conclude, "You are the cause of my stress."

"Knowing others is intelligence; knowing yourself is true wisdom. Mastering others is strength; mastering yourself is true power."

—Chinese philosopher Lao-tze

In my experience, the people who most irritate me usually represent qualities that counter my view of myself: if I view myself as productive, I have a hard time with people I label as unproductive; if I view myself as spiritual, all those non-spiritual people are who get to me; if I view myself as patient, I have a hard time with those I view as impatient. Instead of looking inwardly at what qualities I need to incorporate in myself, I think the answer is to rid myself of such people. Though it seems like a reasonable strategy, life rarely lets me get away with this. No sooner do I manage to push one impatient person out of my life than someone else with this same quality appears in her place. As long as I think, *For me to be happy, that person has to change,* life just keeps sending such people.

Our other choice is expressed in the following Zen story:

A Zen teacher once walked up to a chalkboard and drew a straight line. He then sat down and asked his students to tell him the best way to make the line shorter. One student said to erase some of the line at the top. Another suggested erasing some at the bottom. A third suggested he erase part of the line in the middle. The master shook his head after each response. The students were confused, wondering, How else does one make that line shorter? Believing they had used up all their options, one student said, "We give up. We see no other possibilities."

Walking back up to the chalkboard, the master drew an even longer line next to the first one. "This is how you make the first line shorter," he said.

The lesson of this story, in terms of relationships, is to focus less externally on how others should change and more internally on how we can build our own strengths. I often relied on this when I taught stress-reduction programs to incarcerated teens in New York

City. Though most kids were generally respectful, there were often a couple in any class who either just did not like me or whose challenges were so great that they knew no other way of relating than to battle me. These kids resisted every effort I made: if I asked them to stand, they would sit; if I asked them to sit, they would stand; if I asked them for their opinion, they would be quiet; if I asked them to be quiet, they would talk. For whatever reason, they had determined that, in their words, my "skinny white ass" should not be there. They would do all they could to break me—and they were good.

Though I occasionally had to kick a kid out of class, I realized there were two ways I could view such kids: as mistakes of the universe who should not be in the room, or as ruthless Zen masters out to help me deepen my patience and equanimity. Essentially, I could get pissed off at them for acting how they did, or I could draw a longer line and build my strengths. The former focused on trying to change them; the latter on seeing how I could grow and learn. Curiously, the more I saw the opportunities to strengthen qualities I wanted to develop, such as compassion and equanimity, often the more the students changed. The less I needed them to change, the more they did.

This does not mean we will all of a sudden like such people or that we will no longer be annoyed by their incessant demands, jabs, or complaints. It does mean, however, that we will no longer view them as mistakes of the universe, and in the process we will spend less time judging them. Otherwise, though we interact with someone for only ten minutes, the impact can last throughout the day as we carry the person with us, much like the scenario in the following story:

One day, two monks came to a stream and noticed a finely dressed woman needing to cross. The elder of the two monks offered to carry her over so she wouldn't get wet, and the woman accepted his offer. Afterwards, she thanked the monk, and the two monks went on their way.

A little while later, the younger monk asked the elder, "Why did you carry that woman across the stream? It has been bothering me since we left her. As monks, we are supposed to keep distance from women."

The elder responded, "I left the woman at the river's edge. It seems that you are still carrying her."

Of course, we may decide to physically get away from certain people and exclude them from our lives, but if we no longer view them as mistakes of the universe, then we will not carry them. Otherwise, though their body is not present in our life, their still energy is. I was talking to a friend recently who was telling me with great emotion how a person had been rude to her and how much this person had upset her. From the intensity in which she told it, I would have guessed the event had happened that morning. I later learned that it took place twelve years prior and she had rarely seen him since. The person who upset her was no longer physically in her life, but his energy was; she was still carrying him.

In this tool, we notice the irritation we feel, but we leave the person at the river, or business meeting, or online forum. We focus less on how others could or should change and more on how we can. We draw a longer line, focusing on what we can do, since we are the only user we can control.

APPLY TO LIFE

The next time you are irritated by someone, try this:

- Inquire as to what teaching this person may be bringing into your life. If he was a Zen master instead of a mistake of the universe, how might you relate to him? (Of course, if someone is dangerous, take the appropriate measures. The person could be one Zen master you need to avoid.)
- See what qualities he represents that you find irritating. Is it bossiness, insensitivity, arrogance? Notice if these are qualities that exist in you as well. Look inward and focus less on how he can and should change and more on what lessons you can learn. Explore what it may mean to draw a longer line and build your strengths in this situation.

The Squeeze

Because how we leave one moment is
how we enter the next, it helps to expand instead
of squeeze during times of transition.

If there is one thing that those of us who are constantly connected try to do quite often, it is to accomplish as much as we can in the shortest amount of time. However, if we are not careful, the more we try to squeeze into life, the more squeezed we feel. How many times, for example, have you had ten minutes before you needed to leave somewhere—an in-person meeting with your manager or professor or a video conference call—and you decided to accomplish as much as possible in that precious time? You made a phone call to a customer or colleague that you should have allowed thirty minutes for, started to compose a lengthy e-mail, or began work on an extensive project that took substantial time. You knew such actions required more time than you had, but you figured that with the right effort, you could accomplish in ten minutes what it takes the average human to do in thirty. Essentially, you squeezed.

As a result, you had to end a conversation just as you got to the important matters, or got so immersed in composing an e-mail or working on a project that you left fifteen minutes late to your next meeting, causing you to rush like mad to your next appointment. You sped dangerously through traffic or hurried down the hall and showed up to the meeting frazzled and unfocused, apologizing for your tardiness.

Those ten minutes created a slew of stress that perhaps took hours to run its course. Sadly, none of it was necessary. This happens when we do not give attention to our transition and cannot let ten minutes pass without some Herculean effort to accomplish a substantial task. It is when we feel we just have to squeeze.

Conventional wisdom holds that before a meeting with our manager, what most matters is that we show up on time with relevant information. We need our report clearly written and our supporting material to back it up. While this is certainly important, another often overlooked factor is our state of mind. If we show up to the meeting agitated and rushed, we will be off just a beat, not able to think or act as well. If our transition from what we were doing beforehand to the meeting was stressful, there is a greater chance that the meeting will be the same. Thus, the more we squeeze, the more squeezed we feel.

If we don't squeeze, we may ask, *what do we do with these short transitions*? One option is to do just the opposite and expand in transitions. When transitioning from one event to another, we do it consciously, inquiring what we need to make the transition as smooth as possible. Instead of asking, "How much can I get done in the next ten minutes?" we ask, "How do I skillfully finish what I am doing so I can enter the next event with clarity and focus?" Essentially, we honor and give attention to the transition. We know that the level of ease or tension in which we make the transition will impact the next event.

According to a survey by Pew Research Center, about a quarter (23 percent) of all adults in this country say they "always feel rushed."[18]

How this will look, of course, depends on the situation and what unique ways we discover to clear and stabilize our mind. We may need to review our notes or spend that time relaxing by writing in a journal, sitting in silence, or listening to music. What is needed depends both on the situation and our interests. While the ways we do this may differ, the essence of expanding emphasizes the quality of our mental state as we finish one task and begin another. In this tool, we see that what determines the quality of any interaction is often in the transition; it is to the degree we squeeze or expand.

APPLY TO LIFE

The next situation in which you have a little time before you need to leave somewhere and you notice the squeeze coming, try the following:

- Feel the desire to get as much done as possible right before needing to leave. You may want to pick up your cell and make a call or to start composing an e-mail. Know that you have a choice in how you transition from one event to another.
- Ask yourself, "What can I do so I can enter the next event with the most open and creative mind?" Do what is necessary to expand in the transition so you can show up to the next event with an open and relaxed state of mind. Notice if this affects the next event.

TOOL #13: RESIST EMERGENCY PILOT

Emergency Pilot

The art of nonjudgmental nonparticipation
can help lessen the impact of viewing the world
as if there is always an emergency.

In our age of constant communication, it is easy once we receive certain information to act as if there is a terrible emergency that needs an immediate and emotional response. People often approach us in this mode, wanting us to share their stress and view of a particular situation. Consider the following:

You are at your desk working calmly and effectively when a colleague approaches you with what she believes deserves a serious reaction.

She says to you:

- *"Our boss did not get the report you sent. This is horrible!"*

- *"My friend just broke up with her boyfriend. I can't believe it."*
- *"I can't access my e-mail. My God."*
- *"Google just acquired our competitor! What on earth will we do now?"*

No matter the details, the energy behind the person's words express, "*This* is definitely worth freaking about." Though there has always been a tendency to make our way through the day on autopilot, a common mode for many of us who live constantly connected is to consistently enter emergency pilot, believing that whatever is happening deserves an immediate and emotional response.

In one study, two-thirds of 228 senior executives and managers who responded to a survey say e-mail is the most prominent workplace disruption, followed by crisis of the day (42 percent), personal interruptions (31 percent), and changing priorities (30 percent).[19]

When people are caught in this, they often seek others to join them. As such, they approach us with the attitude, "Can't you see this is an emergency? How dare you look so calm!" Though they may use different words, the feeling is, "You must stress out with me so we can prevent further difficulty from happening! Anyone in his right mind should share my perception of this situation." When in this mode, we believe that whatever is occurring should not be, and only through freaking out, the thinking goes, can order be restored.

We have several options in responding to such a situation. One is to enter emergency pilot along with our colleague. Essentially, we accept and share her view of the situation. We reply, "Oh my God. You're right. Of course you should be freaking out. I will too. Let's do something quickly. This is horrible." Now there are two people in this mode, and the possibility that any effective action will arise is unlikely. More often than not, we expend a great deal of energy, cause more tension to ourselves and others, and have little to show for it.

A second approach is to resist the person's view. We think (and at times say), "Jeez. Just chill out, already. It's not such a big deal. Take a freaking breath and get a grip." In this approach, we try to calm the person down by arguing that she is getting worked up for no reason. The issue is not an emergency, we argue, and we give our reasons why our colleague should not be responding the way she is.

In these first two responses, we either take on the person's view or judge her for it. Yet a third way is to see that the person is simply immersed in emergency pilot and that we need not take a stance for or against her. Exactly how we do this and the words we use depend on the situation. What matters more than our words is where they are directed. Do we speak to her stress by reacting to her? Or do we speak to the underlying issue?

In most situations, the person is seeking help with an issue or she would not have approached us. When we react to her frantic energy by saying, "Chill out," she'll likely interpret it as, "You should not be concerned about what you are concerned about." A person in this state, of course, will not accept this since, to her, the issue is valid. Our telling her to calm down generally only makes her more riled up, sinking her deeper into emergency pilot.

If we neither believe nor judge her view, however, we create an optimal environment for her to let go of this pattern, if she is so inclined, as nothing in us is reinforcing her view by joining her or pushing against her. We are not saying she is right or wrong; we are simply noticing what is true: she is immersed in a certain state of mind. Emergency pilot is often supported by people either joining or resisting. By doing neither, we make room for other options. As such, we increase the likelihood that a more effective and creative mind state can emerge.

Of course, there may be situations that could be called an emergency—a child is about to run through traffic, a fire breaks out in our kitchen, someone is pointing a gun to our head, or our Internet connection is down :-), but in general, most so-called emergencies in our daily life are not worth the stress we suffer by shifting into emergency pilot. Even in actual emergencies, we are likely to better respond from calm and equanimity than from anxiety and fear.

In this tool, when approached by someone caught in emergency pilot, instead of joining or judging, we address the underlying need. We practice the art of nonjudgmental nonparticipation.

APPLY TO LIFE

The next time someone approaches you in emergency pilot, try this:

- See if an action needs to be taken immediately, and if so, do it as skillfully as you can. If a child is about to run into traffic, for example, grab the child. However, if the situation does not call for immediate action, first, take a couple breaths.
- Notice the tendency to either enter emergency pilot with the person or to tell him to "shut up and chill out." The person may look at you thinking, *How dare you look so calm. Freak out with me!* Know you have a choice and can decide not to join him.

Look beyond all the hysteria for the issue that may need to be addressed. Show concern for the underlying issue without joining the person in emergency pilot.

The True Price of Our Two Cents

What we do not say in a situation is often as important as what we say.

In today's world, we can add our view or comment to just about anything—on books at Amazon, on videos at YouTube, and on almost every blog, podcast, and online article. In fact, we can even comment on other people's comments! While it is great such opportunities exist, the challenge is often knowing when not to comment, when not to get in long back-and-forths either in person or online in social networks and forums.

I realized the cost of such comments some time ago when I criticized the political views of a previous commenter in a social network. I knew expressing my view would not likely serve the conversation since the post had no real purpose other than to express my opinion that the person was an idiot. The post set forth an avalanche of responses, some chiding me for my comment, others defending my position. As a result, I spent much of the day consumed in the back-and-forth, which had no benefit that I could see. The exchange, in fact, took days to run its course, as none of us were willing to let the conversation die; we all felt we had to add our two cents and get the last word in. Hours of my workweek were spent dealing with it. If only I had resisted, I would have saved myself a great deal of time and hassle. Those two cents I added ended up costing me a great deal.

As in my case, the comments we make can often set off a chain reaction, much like what the Zen students experienced in the following story:

One day four Zen students vowed to sit in a room and spend the entire day in silence. As evening approached, the temperature dropped, and one of the students could not resist any longer.

"Boy, it sure is cold in here," the first student remarked.

"Oh no, you broke your vow by speaking. Remember, we are not supposed to speak," replied the second.

The third student, on hearing this, piped in, "You dummies. You both just broke the vow because you both talked."

To which the fourth student proudly announced, "All of you messed up. I am the only one who has kept his vow and not spoken."

Like these students' conversation, our comments on other people's views and actions can easily create a similar chain of events. Our "that was not smart" or "I wouldn't have done that if I were you" or "you spoke when you shouldn't have" to a colleague or online commentator begins an exchange that usually serves no one. Often, we do so in order to correct someone after she has made a mistake. More often than not, however, the person already knows this. We are not telling her anything new; we do so for our own satisfaction and feeling of superiority. By taking such action, however, we set forth a judgment ball that bounces back and forth, as the person who is criticized feels the need to respond in the same manner. It all started because we had to add our two cents and did not see at the time the true price of those two cents.

The difference between skillful comments that aid a situation and those that lead to frustration and discord is often determined by the level of judgment present. When we are caught in what we may call the judging mind, we continually look for people and actions to criticize. Instead of a critique that seeks to help, we do so to build up our own sense of superiority. Behind our comment, "That was not smart," is often, "Look how much better I am than you." In fact, in the judging mind, the more mistakes we find in others, the better we feel about ourselves, believing, "Look how screwed up everyone else is and how much better I am." However, this *better* that we feel at such times is actually a hollowness based on fear—a fear that we may make similar mistakes, that we too could be judged.

Recent studies indicate that much can be lost in electronic communication, specifically e-mail, because recipients "hear" the information differently. People think they are communicating more effectively than they actually are, leading to more incidents of misunderstanding.[20]

In this tool, we see the real price of adding our two cents and are no longer lured by the momentary rush of righteousness in criticizing another. Unlike the Zen students in the story, we see the danger of conversations that lead nowhere. Instead, we trust what need not be said.

APPLY TO LIFE

The next time you want to comment without a real purpose, to tell someone, "That was not smart," or, "I wouldn't have done that if I were you," try the following:

- Take a deep breath and feel the motivation behind the action you want to take. See if it is coming from the judging mind, from an attempt to build up your own self-image, or if it may be useful. Consider if it is best to trust what need not be said.

In times of stress, the body is often disregarded. The nights I was up at two a.m. posting on social networks, surfing the Web until I could no longer keep my eyes open, and checking my e-mail every five minutes, I had very little awareness of how my body felt—the level of ease or tension in my shoulders, jaw, or belly. In fact, most of the time I could care less about the state of my body: I simply wanted to consume more information, work harder, and get more done, regardless of what my body had to say on the matter. I saw it as a very minor shareholder in the company whose primary purpose was to park my brain in front of the computer and direct my hands to type, scroll, and click. I lived, for the most part, from my shoulders up. I knew this was not the best way, but I had become hooked.

However, in this pattern my body became tenser and more lethargic, and it was more unwilling to perform as needed. It was not satisfied with the role I had given it: to move my brain around, but otherwise to leave me the hell alone. I soon realized that I was ignoring it to my own peril and that in any battle with it, it would come out the victor. It was, in fact, not a minor but a very major shareholder in the company.

For those of us who are constantly connected, we can easily treat the body as something of a nuisance. When it calls out for attention during the day through a tight belly, burning eyes, or sore neck, we reply, "Leave me alone. Can't you see I'm trying to work? I don't have time for you." We respond to it as if we are our mind, and the body is simply extra luggage we have to carry around; it holds very little significance. The body, of course, is more than this. It can be an incredible learning laboratory to help guide us to a more balanced and less stressful life, but only if we give it due attention, only if we treat it as a major shareholder.

According to one study, pain costs American workers more than sixty-one billion dollars a year in lost productive time.[21]

In this daily practice, we notice how we carry our body, and we begin to soften or relax tension as it arises. In fact, try this right now. Take a moment to tune in to your body. Feel your shoulders, your belly, and your jaw. Likely, you will notice some tension. First, allow it to be. It is the truth of this moment and as such deserves attention and respect. Then without force, soften around the tension. If you are not sure what it means to soften, that is OK. Simply ask the question, What would it mean to add softness to this part of my body? Bring a gentle, kind attention to this place.

In this practice, we are not telling our tension to go away, but are instead adding softness and relaxation around it. Just like taking three full, conscious breaths, the process is to invite, not force. We invite the shoulders to relax, the belly to be at ease, and the jaw to release. We welcome softness and, in doing so, the body naturally responds. This is no different than any living system. If we give kind attention to children or animals, they usually respond with greater energy and life force. The body is the same.

While doing this, if you find yourself thinking, *This softening is not working because the tension is still there,* take a deep breath. The point is to recondition a relationship of ease, and we can only do this by allowing the tension and at the same time inviting softness. If physical tension immediately reduces, great, but if it does not, our intention to soften shifts our relationship to it. Our mind relaxes even if some tension still exists in our body.

This practice helps us to address stress as it arises. We can view stress like water poured into a cup. If too much is added, it eventually overflows, and only then does it get our attention. In the same way, we often only notice stress when it starts to overflow into our life. Only then do we visit the spa or the hot tub or whatever we do. While such activities are useful, we can also soften and release tension as it arises. We can pour out water as it enters. We can soften around the tension that emerges when our computer is taking a long time to boot, when we cannot find the information online that we seek, or when our cell is not getting reception. Instead of reacting in such moments by tightening our belly, jaw, and shoulders, we soften.

In this daily practice, check in with your body at times during the day, and if you notice tension—your shoulders are raised and strained, your jaw clenched, your belly tight—invite softness. If you want to expand this practice further, spend five minutes a day doing the following exercise:

1. Find a quiet place where you can sit upright or lie comfortably on your back. It could be on a bed or outside under a tree.

2. Take a couple breaths. Notice the overall state of your body, if it is tense or relaxed.

3. Feel the various parts of your body, starting with your head and moving down to your neck, shoulders, arms, chest, belly, legs, and feet. Scan your body with a gentle, kind awareness, as if you are an explorer on a new planet.

4. As you do, invite softness to each area you visit. Bring attention to your eyes, and invite softness; to your jaw, and invite softness; to your belly, and invite softness. If one place particularly grabs your attention, you can keep focused on it while continuing to soften.

PART 3

THE BROWSER: CREATIVE MIND

Appearing as if we are making progress on a project is easy. Our hands could be typing and our body moving, but often our activity is not inspired or creative; we are simply going through the motions. Though we are expending a great amount of energy, we have very little to show for it.

A third aspect of Wisdom 2.0 is to focus not on activity for activity's sake, but on inspired or creative activity. The quantity of time we spend looking at the screen or working on a project is not what matters, but the quality of that time—our ability to access what I call the creative, referring to the creative mind. This is the place of both focus and ease where we can see the right action in a given moment, whether we are designing a Web site, writing a report, or relating to a difficult client.

TOOL #15: ACCESS WITH PRESENCE

Better Than Trying

We can often more effectively engage the creative mind through attention than through thinking.

I have learned from various traditions, but I have a particular fondness (you may have noticed) for the teachings of the Buddha and the Zen stories that later emerged from this tradition. In them, you often see phrases that, on the surface, seem paradoxical, such as:

To know Zen, don't know Zen.
To be the Buddha, forget the Buddha.
To get enlightened, refuse enlightenment.

These are the kinds of statements that make you shake your head and wonder if the Zen masters who uttered them were not either senile or consuming too much sake one night. The point of such statements, however, is that by trying to do anything, we can easily create a counterforce. This is why the third Zen patriarch said, "When you try to stop activity to achieve passivity your very effort fills you with activity." In trying to attain stillness, we create more activity; and in trying to be creative, we are often less creative.

This may seem like a strange way to begin a section on engaging the creative mind. You may be thinking, *If I don't try to be creative, what do I do?*

Some years ago, I explored this subject in an interview with George Mumford, who could best be described as a teacher of the zone. He taught regular programs for Michael Jordan and the Chicago Bulls in their championship era, and for the Los Angeles Lakers when coach Phil Jackson moved there. He works with professional sports teams and Olympic athletes on their mental approach to sports, how to make them more likely to be in the zone during competition. The zone is the creative mind, the place of ease and effectiveness where the player knows the right action to take in any given moment. The person is in the flow of the game and her play shows it, making difficult shots or passes look easy. Such a player is not necessarily physically stronger or more skilled than her opponents; instead, her mental state is what makes the difference. This is the focus of Mumford's work.

Of course, more than athletes benefit from entering the zone. We have all had times when we noticed ourselves in the flow at work or school. We were able to see just what a project needed to move forward, we had a wave of creative ideas, or we related to an angry customer with poise and directness. At these times, there was an effortlessness in our action; the right response simply emerged without strain or force. In fact, you may not have even felt that you did anything. It was as if life or the creative was moving through you.

> **"When players practice what is known as mindfulness—paying attention to what's actually happening—not only do they play better and win more, they also become more attuned to each other."[22]**
>
> **—Phil Jackson, who has won nine NBA titles as a coach and is the NBA's career leader in playoff victories and playoff winning percentage**

So, how do we get in the zone? you may ask. I posed the same question to Mumford. He said in his work with some of the top athletes of the world, he often begins, much like the Zen masters of old, by telling them that "if they try to get in the zone, they can't." This may sound odd, but he went on to explain that while effort and practice are certainly needed in any endeavor, accessing the zone can never be ordered or forced. It must come through something else. The something else he offers athletes is, "If they pay attention, the zone will happen as a by-product." This attention, or attunement, is the secret.

This attention is different than thinking about something. If a basketball player is on the court thinking, *I need to get in the zone. I must be in it. Where is that zone?* he is unlikely to perform well, as such thinking increases mental activity. If, however, the player pays attention to his direct experience while on the court—the touch of the ball, the movement of his body, the sounds, and the colors and shapes present—he has a much better chance of playing well. As the player is attuned to his senses, this flow is much more likely.

The same is true for us. No amount of thinking about the creative before we turn on our computer to work will bring it forth. It's not useful to approach our computer thinking, *OK, I'm going to be creative. I'm going to do an awesome job on this and finish it in half the time it usually takes. I am ready. I'm going to do it.* This simply creates greater mental activity and pressure, making it more difficult to access the creative mind. If, however, we sit down at our computer with an open awareness, attentive to our direct experience, the zone is not guaranteed, but it is more likely.

We can see this as the difference between *trying* and *inviting*. For example, imagine that your manager asks you to take on a project with significant responsibility. She wants you to create the next major product for the company, to pitch a highly valuable potential client, to create a new company Web site, or to give a presentation to the board of directors. No matter the details, the project is more responsibility than you are accustomed to having and is an excellent opportunity to show your skills.

There are a number of ways to approach this. One is to be motivated by fear and to *try* to make something happen very quickly. As such, once you sit down at your computer, you think, *I have to do this well. This is my chance. I need to do this project both creatively and effectively. I really need to make something happen with this. Better get started now.* You quickly put ideas down, but they come from force instead of ease.

A second way is to be present and *invite* the creative. In this, you enter the project paying attention to your body and mind as well as to your surroundings. There is less a desire to *make* the creative appear and more a sense of welcoming and inviting. Your breath deepens and your mind opens. This is not something you must figure out but a mystery to explore. The creative is not outside you, something you must struggle to capture or control, but resides within you. You trust that the great ideas you seek are, in a way, already present. Your job is simply to access them.

This inviting can also be useful in challenging situations. When a customer is yelling at us, our manager is criticizing us, or an online commentator is chiding us, we can think, *Why are they doing this? I must respond creatively. I must find a good way to reply.* Or we can look inward, pay attention to our mind and body, and inquire, *What creative response wants to arise in this situation?* The first approach demands and creates more pressure and mental chatter; the second invites, making more room for clarity and ease of mind. It needs to be mentioned that such inviting can exist along with strong emotions. If we are just about to start a big project, fear may be present; if our manager is criticizing us, anger might. We can allow these to be and at the same time look deeper, staying open to the possibility that a creative action will emerge.

In this tool, we approach our work with a heightened awareness, seeing that through awareness and presence, we can access the creative mind and enter the zone.

APPLY TO LIFE

The next time you are about to work on a project that is very important to you, try this:

- Spend five minutes before you work settling your mind and body. Go for a short walk, listen to some relaxing music, or take time to breathe consciously. Find the place in you that is clear and focused. The thoughts, *I need to do this well. I better do this well,* may still arise. That is fine. Let them come and go without pushing away or holding on to them.

- When you approach the task, bring a sense of inviting and allowing, trusting that the ideas you need will emerge as your attention opens. Bring presence to your sitting in front of your computer. Feel your feet on the ground, the pressure of your buttocks on the chair, and the feeling of your fingers touching the keyboard. Work from this presence.

TOOL #16: TRY SOFTER

Hard Versus Soft Effort

It is not *that* we put forth effort that matters most, but *how* we put forth effort.

We have all heard stories about start-ups in which the founders worked fourteen-hour days for several years, then got acquired by Microsoft or Google and netted millions. From their unrelenting work ethic, the story goes, they were successful and all lived happily ever after, never having to worry again. The media, of course, love to tell these stories: Two Young Techies Create Web Site That Is Bought for $50 Million One Year After Launching. These stories are repeated like religious chants in technology circles.

There are other stories, however, not as frequently told, such as the hundreds of other start-ups where the founders worked just as hard but did not get the big payday. There is also the story of the impact of such strained, nonstop, work-all-the-time lifestyles. We do not see many headlines with the title, Two Young Techies Develop Severe Sleep Disorder in Two Years of Work. We rarely hear about the shadow side of such effort.

Perhaps sleep disorders, back problems, exhaustion, headaches, mental fatigue, and ulcers from such a lifestyle are worth it to us. We think, *For a couple million dollars, I will gladly take that*. If we do get the big payday (which statistically, of course, is unlikely) we may be able to justify it. However, no matter the results, we are left with an increasingly bruised and beaten mind and body that we must live with the rest of our life. Even if we made millions, no amount of money can buy us a truly good night's sleep or peace of mind.

Though there is no guarantee that any project will be monetarily successful, what if there was a way to put forth effort that both gave us the greatest chance of success and had the least negative impact on our body and mind? What if instead of having a forced, work-till-you-drop effort with little concern for our mental or physical state, we had an easy, focused effort that was aligned with and beneficial to our well-being? What if we could find the right balance of effort?

People at times resist this, thinking strained effort is all that is needed—that for our Web site to succeed, we must clock massive hours; for our business to grow, we must make hundreds of sales calls; for our friendships to prosper, we must send people numerous text messages and e-mails. Of course, through such strained activity, we may make some progress, but it comes at a hefty, and usually completely unnecessary, price.

> **"The moment one constructs a device to carry into practice a crude idea, he finds himself unavoidably engrossed with the details and defects of the apparatus. As he goes on improving and reconstructing, his force of concentration diminishes and he loses sight of the great underlying principle. Results may be obtained, but always at the sacrifice of quality. My method is different. I do not rush into actual work. When I get an idea, I start at once building it up in my imagination. I change the construction, make improvements and operate the device in my mind."[23]**
>
> **—Nikola Tesla, legendary inventor**

We can be much like the martial arts student in the following story:

Once a young man in Japan decided he wanted to be the greatest martial artist in all the land. He thought that to reach this goal, he must study with the best instructor, who lived many miles away. One day he left home to do so. After traveling for several days, he arrived at the school and was given an audience with the teacher.

"What do you wish to learn from me?" the master asked.

"I want you to teach me your art and help me become the best martial artist in the country," the young man replied. "How long must I study?"

"Ten years at least," the master answered.

The guy thought, Ten years is a lot of time. I want to get this done sooner than that. Certainly, if I try harder I could do so. He then asked the master, "What if I studied twice as hard as everyone else? How long would it take then?"

"Then it would take twenty years," replied the master.

The student thought, That's even longer! I don't want to spend twenty years learning something. I've got other things to do with my life. He must not understand just how hard I will work. So the student inquired again, "What if I practiced day and night with all my effort? Then how long would it take?"

"Thirty years," was the master's response.

The young student was confused and wondered why the master kept telling him it would take longer. In desperation, he asked the master, "Why is it that each time I say I will try harder, you tell me it will take longer?"

"The answer is simple," the master replied. "With one eye focused on your destination, there is only one eye left with which to find the way."

Essentially, the teacher was telling the student to try softer, that effort alone would not help him reach his goal, and that he needed the *right kind* of effort. You probably do not want to become the best martial artist in the land, but you may wish to be the greatest coder, Web site designer, or Internet entrepreneur. No matter your goal, the same teaching applies.

Of course, no amount of sitting around wishful thinking is going to write our report, build our Web site, or grow our business. It takes effort, resolve, and commitment. But like the martial arts teacher in the story, if you are like me, you have seen projects fail not because of insufficient effort, but due to a lack of attention to the process. The quality of the effort was what made it more difficult for people to get along and work creatively. The fear and tension, not a lack of effort, were what increased the likelihood of the project's failure.

The advice from the martial arts teacher is pertinent to us as well: we can often make more progress and with less stress not by trying *harder* but by trying *softer.* By doing so, there is an ease to our effort, a stillness to our actions, a patience in our work. We focus not just on the *what* that we hope to accomplish, the goal we are trying to reach, but on *how* we work toward it—whether our two eyes focus only on the result or are also attentive to the process. You can see this in great athletes and performers; they have a beauty and ease to their actions, making everything they do look almost effortless. We can bring this same quality to our work—be it coding, writing, or designing—and usually make greater progress as a result while taking on less physical and mental tension.

APPLY TO LIFE

The next time you are working very hard on a project and are frustrated at the lack of progress, try the following:

- Take a step back from the work. Relax your focus. As you do, expect some resistance. Your mind will likely respond, *Don't halt. Keep going! If you stop, everything will fall apart.* See the conditioning based on the notion that effort alone matters.
- Explore a softer approach to the endeavor, one that can see the situation unclouded by force or struggle. Let your breath deepen, your eyes relax, and your vision expand.
- Next, engage in the project again, paying attention to the quality of your effort. Know that you will more likely succeed through a focused and relaxed effort rather than a tense one. Stay open to the possibility that you can complete the task at hand without the extra stress.

TOOL #17: BEFRIEND SILENCE

Slow the Train

Both our greatest fears and our most creative ideas reside in silence. We both want and fear it.

Silence is a rare and often unsettling experience for many of us in the technological age. When the television is not on, the computer is off, our iPod is not playing, and no one is around to speak with, we often more distinctly feel our anxiety and frustration. We hear our ever-active chattering mind creating a laundry list of to-dos or carrying on multiple conversations with people not there. We contact what Eckhart Tolle refers to as "the voice in the head."

Various voices are within us, of course, including creativity and wisdom. But there is also the voice of constant mind chatter, and this is what we often first meet when silent. This contact leads some people to think that silence creates more mind chatter. "If I just keep busy and make sure there is always noise, then I'm fine," they claim. However, silence does not create the chatter; it reveals it. And here's the catch: the mind chatter still affects and directs our life; it still clouds our seeing, even when we can subdue it with noise.

Due to contact with this mind chatter, we often avoid silence—at such times we quickly call someone, surf online with no real focus, turn on the television with no show in mind, or play our iPod with no interest in any particular music. We do anything that takes our attention away from what may emerge in silence. As a result, we engage with technology to avoid rather than to create. We run from instead of move toward life, all because we are uncomfortable in silence, because noise has become a means to keep our fear and anxiety away. In this pattern, silence is an enemy, not a friend. As such, we are never able to see the wisdom and intuition underneath the chatter, the deeper knowledge that can be accessed in silence.

"In the attitude of silence the soul finds the path in a clearer light, and what is elusive and deceptive resolves itself into crystal clearness."

—Mahatma Gandhi

The real benefit of silence, however, is not simply a matter of stopping external noise—though that can help at times—but of allowing our thoughts to settle so all of our senses can open, and we can see thoughts more clearly. This is not so much a matter of changing the *content* of thoughts, but the *speed* at which they come. For example, if you are watching a freight train pass in front of you at one hundred fifty miles per hour, and someone asked you what company's names are on the various railcars, you would not be able to know. You probably would have a hard time even knowing the colors of the railcars. The train is simply going too fast to notice such detail. However, if this same train passed in front of you at five miles per hour, you would know these details. You could more easily read the names and see the colors of each railcar, possibly even noticing the small space between each one. The only change to the train is the speed in which it passes you.

The same is true with the mind. With a mental train speeding by at a hundred miles per hour fueled by continuous noise, it is more difficult to see the creative thought, the out-of-the-box idea, or the unique perspective. Thus, we spend hours in a meeting or looking at a computer screen trying to find an answer that we may be able to discover if we could stop long enough to trust silence, to discover what is already there if we slow down enough to see it. The best way to do this is often to befriend silence.

APPLY TO LIFE

Make space for silence in your life.

If you are accustomed to working with music playing, try doing so without it. If you always play music while you drive, try doing so without it. If you always have your iPod with you when you run, try leaving it at home. If you usually watch TV or movies online before you sleep, try simply lying in your bed in silence.

Gently make friends with such silence. At first your mind will likely resist this and counter, *I'm bored. Give me something to do.* However, as you spend more time in silence, you will be able to touch deeper levels beyond the mind that continually needs to be entertained.

The Creative Body

The creative can be accessed via the body as well as the mind.

Though I often use the phrase *creative mind,* it is more accurate to say creative mind/body, since the two are deeply intertwined. For those of us constantly connected who spend much of our day sitting in a chair looking at a computer screen, it is easy not to incorporate the body into our awareness. Our attention is generally focused outward, as we try to solve problems, fix glitches, and find the right words for an e-mail or a report. We often forget to tune in to and benefit from a knowing of the body. Such bodily wisdom is not new. People have always coined phrases, such as *gut instinct,* that speak to an intuitive knowing felt in the body.

In accessing the wisdom of the body, its shape does not matter—tall or short, thin or fat. Instead, we focus on inhabiting the body with our full awareness, being present in the body. Doing so while engaged with technology, when our attention is usually directed at a screen, can be enormously challenging, yet equally as beneficial. This is not so much about giving attention *at* the body, but *in* the body. We can look at the body, for example, and judge it as pretty or ugly, as fit or unfit, as too skinny or too fat. We can also give the body attention in terms of how best to dress and clothe it. Another way is to *tune in* to the body, to feel it as a living system, to experience the aliveness and pulse of the body. You may have more distinctly experienced this at certain times in your life, possibly when you were out on a jog, walking in the woods, or playing a sport. You could feel the energy and life force of the body.

"All the best things I did at Apple came from (a) not having money, and (b) not having done it before, ever."[24]

—Steve Wozniak, cofounder of Apple Computer

What we experience in a given moment can be noticed as equally in the body as the mind. How many times, for example, have you found yourself completely frustrated that a piece of technology is not working right—your computer keeps freezing or a desktop application is not responding—and you realize that this frustration is also manifest in your body? It is not just that your mind is upset, but your body is as well through a clenched jaw, a tight belly, and a hunched-over posture. We experience it in our mind/body.

The body can be an excellent means to help us know whether or not we are creatively engaged with an effort. Generally, when we have little access to the creative, clues usually show in our posture—our head is often tilted forward with eyes glued to the screen, our shoulders are high, and our breathing is shallow. The creative could find a way in despite this, but the road is much bumpier. On the other hand, in those times when we are engaged creatively, our body generally reflects this with a more upright and balanced posture. Our breath is fuller, eyes softer, and shoulders more relaxed. In fact, if we ever want to know what is directing our efforts, we can often more easily find the answer by tuning in to our body. It will let us know.

The more attuned we are to the body, the sooner we can notice when we are engaged stressfully instead of creatively. Just as shifts in the mind impact the body, so too do shifts in the body affect the mind. One helpful tool is to bring awareness to our body as we engage with technology. Instead of all our attention focused externally, often looking at a screen, we include our body in our awareness. You can try this now. As you read these words, bring attention to your body, including your breath. Instead of leaning forward, trying to read as quickly as you can, gently receive the words, staying tuned to your body as you do. Be present in your body as you take in information.

When the body is engaged in this way, when it is brought into our awareness, we can use it much like a diagnostic tool. When we notice our jaw clenched and belly tight, we are likely not working as effectively as we could. We can then take a breath, soften around the tension, and more easily regain access to the creative mind, to the place and ease and focus where we can most effectively engage with our work.

APPLY TO LIFE

The next time you approach your computer to work, try this:

- Notice the posture you take. What is the placement of your shoulders, your head, and your feet?
- Without judgment, inquire, *If I was going to sit in a way that invited the creative, how would I do so?*
- Experiment with sitting upright in your chair, back straight but not rigid, feet flat on the floor facing forward, shoulders back slightly, head resting comfortably on your shoulders, and belly open and full.
- As you work, keep attuned to your body. Let your work come forth not just from your mind, but from your entire body.

TOOL #19: ALLOW STUCKNESS

To Get Unstuck, Be Stuck

The best way to get unstuck is to accept the moments when we are stuck.

We have all had times when we are working on a project at school or work, and it was initially proceeding quite smoothly, but then it hits a wall. The Web site we are building is at a standstill, the report we are writing is stalled, or the PowerPoint presentation we are working on is making little progress. The clarity and excitement we once had for it is gone, and we are at a loss for what to do. Though we were once making progress, now every effort we make produces unsatisfactory results. We struggle to break through the block, but the harder we try, the more stuck we feel. It is as if a large boulder has been thrown in our way with no easy means to get past it. At such times, we often think, *If I could just get rid of this block, all would be fine.*

In such a situation, what is often the most challenging action—and the most needed—is to acknowledge what is true. The answer, essentially, is to be stuck.

You may be thinking, *Dude, at these times, I know I am stuck. How will acknowledging that I am stuck help at all? I already know!*

Ironically, though it may seem that we are continuously experiencing stuckness, usually what we experience in such situations are moments spent trying to get unstuck. We experience our resistance to what is happening more than what is underlying it. In this sense, we are rarely just stuck; we are often trying to get unstuck. It is similar to the following story a teacher posed to his students about how best to escape from prison.

> *"If you want to get out of prison," a teacher asked his students, "what must you do first?"*
>
> *"Try to bribe a guard and get him to let you out," one person suggested. "I think that would be the first thing I would do to get out of prison."*
>
> *"No," another suggested, "I would get the best lawyer I could find."*
>
> *"The first thing I'd do to get out of prison," offered another, "is make friends with other inmates and see if we can come up with a plan to escape."*
>
> *None of these, however, was the correct answer. According to the philosopher Gurdjieff, the first thing one must do to get out*

of prison is to acknowledge that one is in prison. Before that, no escape is possible.

His point is this: how can you get out of something you don't know you're in? This is also true for us in everyday situations. The basic truth still applies: we cannot change that which we first do not accept. If we are stuck and cannot allow it to be, then we will never—or at least not as quickly—get unstuck. The best way to change what is, is to accept it. This is true no matter our situation, whether we are in prison, writing code at Yahoo! or Google, studying at college, or launching the next great Internet company.

Stuckness is often hard to accept since along with it comes a slew of berating voices, such as, "I'm bad. I should be doing better. Why is this happening to me? I thought I was so much more advanced than this." Amidst this barrage, we often try to distract ourselves—we visit the snack bar, surf online, or watch silly videos—all to push away the truth of the moment, which is stuckness. You see people in this state walking aimlessly through the halls of businesses. If someone asked them, "What are you doing?" the most honest answer would be, "I'm trying to avoid stuckness."

We don't want to avoid stuckness as much as we do the voices that come with it, the judgment we feel in such moments. However, if instead of fighting it, we can just be stuck, the experience is actually not nearly as painful as trying to avoid it. In fact, in seeing that this is simply the truth of the moment, the tension and blame around it subsides.

> **"If you have some plan and it doesn't go that way, roll with it. There's no way to know if it is good or bad until later, if ever."[25]**
>
> **—Evan Williams, cofounder of Blogger and Twitter**

Once we accept a moment of stuckness, an interesting shift can take place: we can more clearly and creatively see the path to get unstuck. In fact, by accepting stuckness, we are already unstuck. We may not yet know how to proceed with a project, but the frustration around the experience subsides. Then we can take action. Before this, our mind was clouded by our resistance. Once that resistance lessens, the clouds disperse, and we can see what is actually there and better know how to respond. By doing so, instead of tension and frustration motivating us, we act from the ease and clarity of the creative mind.

This does not mean we like to be stuck, but that when it happens, we acknowledge it as the truth of the moment. However, if you find yourself saying, "Shit, I'm stuck. OK, I accepted it. Now it better change," that is not acceptance. That's using words to mask resistance. Of course, what is true may be that we do not accept the stuckness—that is fine too, and it gets us one step closer. The important point is to accept what is true, not what we think should be.

Of course, the initial response to such a moment may be, "Shit, I am stuck," but if we let that thought arise and pass without judgment, we can discover a deeper place inside us that responds, "Ah, stuckness, show me your wisdom." We open to the available teaching. Stuckness is experienced as an expansion, not a contraction. It is not "I'm stuck and poor me," but, "Here is this stuckness. Let's see what this brings."

Most successful companies have been in such a place and often had to adapt their plans accordingly. The founders of Google, for example, originally thought they would primarily make money by leasing its search engine to various Web portals, Yahoo! started as a way for the founders to share their favorite Web sites with friends, and Starbucks began as a place to sell coffee beans for people to take home and make their own coffee. In the growth of these companies, they had the choice to either fight the new information they were receiving or to accept and flow with it. Whether it is personal or at a company, stuckness at times offers a new direction waiting to be explored. We can more easily find this when we switch from thinking, *I shouldn't be stuck; this shouldn't be happening,* to an interest in what the new circumstances provide.

This can only be discovered, however, when we wholeheartedly accept moments of stuckness. In this way, the best way to get unstuck is often to be stuck.

APPLY TO LIFE

The next time you find yourself stuck while working on a project, try this:

- Notice the tendency to personalize it, thinking, *Why am I stuck? How could I get stuck? Why is this happening to me?* Instead, simply note, "A feeling of stuckness is present." Let it be. It's OK. Feel and allow it. Acknowledge it as the truth of the moment. Know that the first step in escaping from prison is to know you are in prison, and the first step to getting out of stuckness is to acknowledge that you are stuck.
- Next, open to learning from this situation. Inquire, "How might I creatively respond to what is present?" Stay open to learning from the block rather than trying to quickly get rid of it. You do not have to try to be happy that the block exists, but know that because you have accepted the truth of the moment, learning is possible.

We have all had days when our work seems to increase even as we diligently try to keep up. For every item we cross off, two more appear in its place. On these days we often think, *Man, do I need a break*. However, when we are finally able to take one, we spend it in such a way that when we return to work, it did not feel like a break at all. In fact, our long-awaited break added stress because we spent it in the same mode in which we had been working: we made several phone calls, got involved in a debate online, or spent the time complaining about work to one of our colleagues. We may have been better off, in fact, had we not taken a break at all.

Since most of our days are filled with activity—phones ring, e-mails arrive, reports and papers must be written, Web sites explored—breaks are best if they take us away from such actions. Yes, we can spend our break surfing the Web or playing Halo, but we are likely to return to our work with even less energy. Since many of us spend much of our day looking at screens and communicating, we need breaks that do just the opposite, that reduce the amount of information we digest and that give our attention a break.

Such conscious or real breaks are often not supported or encouraged in work environments. Our manager is not likely to ask us, "Have you recently been taking breaks that aid your mind and body?" Neither, often, are our colleagues. In fact, both may encourage us to do the opposite by asking us work-related questions while we are on our break. Such support will likely emerge in the business of the future where the quality of one's mind and body at work is given greater attention, but until then we must take a certain resolve to best use our breaks because no one else will ensure it for us.

Before I offer some ways to make the best use of breaks, let me define the two breaks in most of our workdays: arranged and integrated. What I call arranged are those that are primarily set in a day, such as the thirty minutes or hour for a lunch break that we can spend at our discretion and the fifteen-minute breaks we are generally allowed in the middle of the morning and afternoon. These we may call expected or allowed breaks when we are away from our phone and computer.

The second, what I call integrated breaks, are those that we can take while at our workstation. We incorporate them during our day. These are generally much shorter, lasting only a minute or two. They include the first daily practice of taking three conscious breaths at various times in our day. At such times, our body is still sitting in our chair waiting for our computer to boot, but we are using it internally as a break, as a time to focus and calm. Most of us already know that such integrated breaks are useful; the challenge is remembering to take them. One tool in this regard is to download a mindfulness bell onto your desktop.[26] You can set it to ring at various times in the day. I have mine set to ring every hour. This bell can serve as a reminder to take a minibreak each hour—even if it is just a few breaths.

In one study measuring people's heart rate recovery to stress, there was a significant decrease when people were exposed to a natural scene they viewed through a window. When that same scene was displayed on a high-definition screen, the recovery response was much less. It had about the same impact as looking at a blank wall.[27]

If you do so, you may notice, as I have, that when the bell rings at the end of the hour, if I am immersed in a project, I will fight the bell's invitation to take a short break. I will first question the bell's authenticity, "Has it really been an hour since it last rang? It can't be. The clock must be off." When I see that the clock is correct, my mind will often argue that I don't have time to take a few conscious breaths—there is some great issue or problem that needs to be solved right away. Though I know deep down that the one minute I spend tuning in internally by taking a few conscious breaths would make me more able to address the issue at hand, still the initial tendency is to resist it, to think that I am so busy and my time so important that I cannot spend one minute every hour to tune internally. Over time, however, as we see the importance of our mental state in our work, we can welcome and apply such mini- or integrated breaks during the day.

In terms of the longer or arranged breaks, here are some tips to make the best use of this time:

1. **Reduce Information Intake**. As mentioned before, during your break take in as little information as possible, including news, games, images, and other information.

2. **Breathe Deeply**. When working, it is often very easy to pay little attention to our body. Our breathing becomes shallow and tight. On breaks, bring attention to the breath, let it help reenergize your body and refresh your mind.

3. **Go Outside**. The world begins to look very narrow when we spend much of our day looking at a screen. The expansiveness and fresh air of the outdoors adds energy to the body and mind. Even if your office is in the heart of Manhattan, there are always signs of nature. We have the sky above, air around, and plants and grass coming up between the sidewalk cracks.

4. **Move**. Throughout most of human evolution, we have lived as mobile beings. Our bodies are not built to sit in one place all day. They get energy through movement. But now that many of us spend much of our days sitting still, during our breaks it helps to move, to go for a walk, or to do some simple yoga or chi-gong exercises. By doing so, when we do return to our chair to work, our body is generally less tight and our mind more focused.

5. **Keep Communication to a Minimum**. There is nothing wrong with communication via phone or computer, but if that is what we do much of the day, on breaks it helps to limit such as much as you can. Keep break times for yourself.

Also, many people exploring a more conscious relationship with technology take the longer breaks we receive in a week, often the two days on the weekend, to take time away from technology, including not checking work e-mail. When they do check work-related e-mail or make work calls, people often say that they feel as if they are still at work on the weekend, processing problems, thinking of responses, and mentally composing reports. Then when they return to work on Monday, it was as if they never had a break. Physically they may have left, but mentally they were still there.

As more people realize this, a growing number are starting to spend their days off doing other activities away from their computer screens or work-related matters, and are discovering, to their great surprise, that they do not miss all that much. The world does not collapse from their not checking e-mail for one or two days. Generally, life goes on just fine. When they are able to not just physically but also mentally take a break from work, they also find that they are able to return to work on Monday more energized, making the days they do work in the week much more productive.

The Future That Is Now

The only way we experience the future is
when it becomes the now.

One way we miss the creative is by putting our attention on a possible or planned future event. While it is natural to look forward to future events, when we fixate on them we have less access to the creative mind. How often, for example, have you heard someone involved in an endeavor talk incessantly about how great their life will be once their project hits it big? She tells anyone who will listen about the money, increased power, and greater acknowledgment she will have once this future occurs. Much of her life is invested in and lived through this future scenario. This is a particular tendency in Internet start-ups. When I see people in this state, I know their efforts have little chance of reaching their imagined future because they have less attention to do the work needed in the present to make that future more likely. Their attention, instead, is off in the future. We might call such fixation on what could be *futuring*. Here is one example:

A woman once went to visit a psychiatrist to get help with her husband, who was a successful advertising executive and salesman.

"What is the problem?" the psychiatrist asked.

"Well, it's our lovemaking," the woman said.

"Yes? What happens?"

"That's the problem. It doesn't happen. My husband just sits at the foot of the bed and tells me how great it will be."

We too can sit at the foot of the bed of life and talk about how wonderful the future will be, whether it is lovemaking or getting that big financial payday. However, in doing so the actual work to help create the conditions—the lovemaking, the building of our Web site, the launching of our company—gets less attention. Futuring does little to help us in the present, which is the only place life is ever lived and the only place we can access the creative mind to accomplish our work.

Futuring is developed and encouraged by the aspect of mind that believes that this moment right now is quite worthless and undeserving of our attention, and that only a future moment—when our Web site succeeds, our YouTube video gets a million views, we become one of the top gamers, or we get the promotion at work—will allow us to experience a moment worthy of our attention. Until that happens, the thinking goes, the future should consume us. We are sold on a vision of the future at the expense of the present.

"The fact that we are fairly engineering-centric has been misinterpreted to mean that somehow the other functions are less important. The fact is that we want everybody in the company to be innovative."[28]

—Sergey Brin, cofounder of Google

When we buy into the future, when we believe its promises, there is no end to our buying. We must continually invest more and more of our attention there since futuring never really ends, even when we get the results we seek. One future is soon replaced by another. We think, *I may have succeeded once, but as they say, "Once you are lucky. Twice you are good." I must achieve more.* This then transforms into, "Twice you are good, but three times you are great. . . . Three times you are great, but four times you are awesome. . . . Four times. . ." In this sense, there is no end to the future.

The difference between futuring and working creatively and passionately toward a goal is our level of investment in the outcome, particularly how much we believe it will increase our value or sense of self, how much it will prove that we are someone who is worthy or successful. In such an orientation, because so much depends on the future, we are weighed down and less likely to work creatively or effectively. When we work passionately, on the other hand, our attention is on the process, on planting the seeds in this moment that have the greatest possibility of creating a positive future. Because our worth is not dependent on the results, we can work more fluidly and easily. We know the best way to create a satisfied future is through a satisfied present, that the best chance of accessing the creative mind at a future moment is to do so in this moment now. In fact, when the future comes, it is just another now. If we are not creatively engaged in the current now of our lives, what makes us think we will be at a future now?

In this tool, we see that our real work is to focus on what we can do right now to engage creatively with our work and life. By doing so, the future will work itself out.

APPLY TO LIFE

The next time you are consumed with a future event, try this:

- Be aware of the experience of looking forward to the event. Notice how you view and orient to the world from this perspective. See if there is any hope that somehow this future event will make you a better person, will prove your worth.
- Explore what it may mean to bring the sense of completeness from this future event into your life now, as if the event you hoped for has already happened. The sense of satisfaction you were hoping to get has already been given to you. Though the external may not have happened, internally it already did.
- Live from this place of wholeness, putting the conditions into place now that can help bring forth the future you would like. However, you do not need the world to change in the same way, since the internal shift has already happened.

Clear Personal History

The present may be challenging, but we make it more difficult when we bring in the past.

The other way our creative energy gets zapped is through living in the past. We have all had difficult moments, times when a project at work was not progressing as we had hoped or our manager was pressuring us. During such times, we have a tendency to add to our difficulty by bringing in the past. We think, *This project was going so much better last week*, or, *My last manager was so much better. Compared to her, my current one is a pain in the ass.* While the moment may be challenging, the greater difficulty comes as we measure it to a time in the past when conditions were different. Essentially, we bring the past into the present, but it generally only makes the present more difficult.

I experienced this when my job for a celebrity ended. When I first started it, I was amazed at how often and how quickly people returned my e-mails and phone calls. All I had to say was that I was calling from the celebrity's office and, more often than not, I soon heard back. I learned how much fame inspired people to communicate. After a year or so in the job, I became accustomed to this. Of course, I knew people were not responding because of me but were doing so because of my relationship with a certain somebody, but I accepted it nonetheless. I was a somebody by association, but I would take that. At least I wasn't a nobody.

However, once I left the position, I was pained to discover that my status as a somebody went with it. I was now just a regular guy, and it sucked. My phone calls and e-mails were not answered with near the speed or consistency, since I could only leave my name on answering machines, and it carried little weight. I was now a nobody. It was as if I had my popular username stolen by a hacker overnight, replaced by one that had no friends or activity history.

As any ambitious person would, I fought this with all my might, desperately seeking a way to recover my somebodyness. I pleaded to the universe, "You have made a mistake. I am a somebody, not a nobody. Please correct this at once!" I brainstormed ways I might recover my somebodyness: "Is there another celebrity I could work for? How do I become important again?"

Much to my initial chagrin, the universe did not heed my calls. After great pain and suffering, I realized I had another choice: instead of trying to recreate the past, I could let it go. I could let my past be history. I saw that no past event was causing me pain; instead, the fact that I was trying to bring my past into my present, to continue a sense of specialness, was. My attachment to the past was making my present excruciatingly painful and limiting my ability to respond creatively to the new conditions. In my case, I could not see that a less busy life that was not centered around people society deemed important actually provided me with the opportunity to focus on other areas of my interest, such as writing.

The nature of our past, whether it was full of success or failure, of pleasure or pain, of happiness or misery, does not matter; if our attention is focused there, we have less ability to engage creatively with our current projects and work. Just as living in the future can drain our attention, so too can living in the past. If the past was very difficult, we may need a good therapist or counselor to help us release it. Often, however, we carry a pleasant past. We had a situation that was deemed worthy by the culture: a job that made a lot of money, popularity on a social network, excellence at sports or academics in high school or college, or partnership in a successful start-up. Nothing is wrong with these experiences, but when conditions change, as they will, it is easy to carry the past with us, to try to keep alive the identity of ourselves as a someone of worth, even when conditions do not support them.

In talking about how he decides who to hire, Larry Brilliant, head of Google.org, said, "I have kind of graduated from looking at the scorecard of degrees and SATs. I am looking for a WAT, a Wisdom Aptitude Test."[29]

The initial tendency when seemingly pleasant conditions change is to try to recreate them. I, for example, sought other famous people to work for in order to sustain the feeling of importance I experienced in my previous work. The remedy, we think, is to freeze life, to not let conditions change. We try to hold on to the past, not drop it. However, in our attempt to freeze life, we become frozen; we cannot see the opportunities in our present situation. No matter how golden the past, the present always holds more potential.

Of course, the past is important to acknowledge. We have learned lessons, faced certain challenges, and achieved particular accomplishments. To make the past history does not mean that we forget or deny our past, but that we no longer carry it; we no longer hold on to the specialness we may have experienced at one time. In Zen, this is referred to as "leaving no traces." We let the past be the past, with no traces.

We all likely know people for whom the past is not history. They act arrogantly due to past successes or hang their heads low because of past failures. Often, they focus only on the positive: they talk about how popular they were in high school or college or how successful their last company was. Much of their life is lived through a past event they view as who they really are. They often introduce themselves as the former holder of a certain position. Of course, if such information is relevant to a conversation, it may need to be mentioned, but if this is done all the time, it is a sign that the past is being carried into the present. People like this are leaving traces.

This is why sometimes those who have had early success have a hard time creating new products or businesses. They think, *I succeeded last time, so I better succeed even more this time.* They carry the past and are not as free to experiment—to try, fail, and learn. Because of their past success and reputation, they are hindered in their current efforts as they worry about how other people will view them if they do not succeed as hoped. Because they cannot fail, they have a harder time succeeding.

The challenge is to make the past history so we can relate to our present life and work with greater ease and creativity. We see that it does not matter how many of our past efforts failed or succeeded, or who we were last year or last week; what matters is how fresh and open we meet our life and our work right now, with room to fail and to succeed.

APPLY TO LIFE

The next time you are working on a project and find yourself thinking, *But in the past . . . ,* acknowledge the past, and then inquire what would happen if you made the past history, if you did not carry it into the present.

Know that you do not have to try to recreate the past, be it pleasant or unpleasant, but that you can instead focus on and trust the current conditions.

Chance Favors a Prepared Mind

We cannot guarantee particular results,
but we can make them more likely.

In my experience, two states of mind often get in the way of any creative endeavor. The first is doubt. Though the specific voice of doubt may be different for each of us, it often carries the thoughts, *I can't do this. I should just give up. It doesn't matter anyway. Whatever I try, I will fail.* When doubt is strong, we lack the energy or confidence to move forward. The future seems predetermined against us. We believe that either life will not change or, if it does, will do so for the worse.

The other mind state is unwavering certainty. This is the state of mind that says, "Life must work out just how I think it should. It must be this certain way. I will accept nothing else. I will succeed no matter what." We have an image of how life should be and seek to mold the world into it. In this mind state, failure is not an option. Though numerous conditions may be outside our control, we believe we can force life to our desires. When we get caught in either of these states, in doubt or in unwavering certainty, we may make some progress, but it's as if we are driving a small car hauling a trailer full of bricks or surfing the Internet on a 56k modem.

A third or middle way is to take actions that increase the likelihood of success while at the same time surrendering the outcome. We don't seek absolute control, thinking, *I am going to do this, no matter what. It is up to me.* Neither do we avoid responsibility, saying to ourselves, *This is completely out of my control, so I may try a little but not much. It doesn't matter anyway since it won't work out.* This third way asks us to do our work with both complete confidence and complete surrender, with deep passion while letting go of any attachment to the results.

One example of this middle way is Google. When cofounder Sergey Brin was asked at the annual Web 2.0 conference of industry leaders about Google's success, he responded, "I think probably the number one factor that has contributed to our success over the past seven years is luck."[30] It was a surprising but honest answer. Though it may seem he was just being humble, there is some truth to this, particularly since the model Google used to monetize search had been used before by Bill Gross, founder of Overture, later bought by Yahoo!. However, Gross made the effort while the Internet was just a little too young, and there was not a significant interest in online advertising. Google, after building a large dedicated user base, launched its now famous AdWords, very similar to Bill Gross's attempt. But this time it hit, big time . . . and the rest, as they say, is history, one that you can read more about in John Battelle's book, *The Search.* Essentially, Google was lucky, Gross was not.

"We don't have much in the way of a business strategy. Like no business plan. . . . The deal is, it's a mixture of luck and persistence."[31]

—Craig Newmark, founder of Craigslist

Google's success, however, has been much more than luck. In spiritual circles, there is the saying, "Enlightenment is an accident; our job is to become accident-prone." An old Chinese proverb makes the same point this way: "Chance favors the prepared mind." While no one can guarantee the success of an effort, we can take actions that make it *more likely* we will do well, whether in starting a company, taking a college class, or practicing a creative art. We can make ourselves more prone to succeeding. So while Google founders Sergey Brin and Larry Page were lucky, they increased their odds by their prepared minds and innovative approach to business, which prioritized creativity by working in small teams that act like Internet start-ups and by giving their engineers what they call 20 percent time, meaning that 20 percent of the person's work in a week can be dedicated to developing products he thinks are valuable, not just on those that are coming from the top down. Essentially, Brin and Page were both lucky and worked hard to be success-prone.

When we find this balance in our life and work, we are no longer overcome by doubt or the need to control an ever-changing world. Instead, we see that because chance favors the prepared mind, we do our part impeccably and at the same time completely surrender the results. When we can do so, we get good at being lucky.

APPLY TO LIFE

In your work, notice if doubt or unwavering certainty is motivating you. Do you doubt you can complete a project, or are you trying to force a particular outcome? Is one of these qualities motivating you more than the other?

Explore what it might be like to approach your project with complete confidence and total surrender. If you have a lot of doubt, know that while you cannot guarantee any results, you can create conditions right now that may make certain results more likely. In each moment there is such an opportunity. If unwavering certainty is present, know that the results are never yours to decide and that you will work more effectively when not holding to them.

In the balance, you will find that often simply the joy of creating and the process of discovery become the motivating factors.

TOOL #23: LIVE, DON'T LOOK, CREATIVE

If You Meet the Creative on the Road, Kill It

Our ideas of what it means to be creative often make us miss it.

Most groups have norms that members follow; certain beliefs and ways of dressing are encouraged. Often, no one will verbalize these, but over time people begin to think and dress very much alike. The same can happen with the creative; we begin to think there is a look of the creative, and then try to fit the part instead of living the experience. We buy the right techie clothes, the best computer, the most sophisticated cell phone, we know all the right people and latest trends, are popular on the best social networks, and yet we have little access to the creative mind. We look and act like we do, but we don't.

This tendency was pointed out some years ago in a book published with the curious title, *If You Meet the Buddha on the Road, Kill Him. Why,* you may ask, *would an author encourage people to kill someone, especially a Buddha? How could someone promote such a violent act?* The book, it turns out, was not written by a religious zealot out to destroy members of other religions. Instead, the author wanted to emphasize the danger of holding beliefs about how one should be. The author was not encouraging people to kill an actual Buddha, but to destroy their *ideas* of what it means to be a Buddha or any other spiritual figure.

Still, you may ask, *Why would someone want to kill her ideas about the Buddha? What's wrong with such ideas?* Nothing is wrong with them, but when we believe our ideas of how something is supposed to look, we get caught in form; in other words, we focus on images instead of direct experience. We act how we think a spiritual person should act instead of trusting our own expression. However, we don't need to kill these ideas so much as we just need not believe them.

The same is true with the creative. It is easy to take on ideas of how a creative person should act or look. We think creative people wear dark turtleneck shirts like Steve Jobs, casual dress shirts like Bill Gates, flip-flops like Facebook founder Mark Zuckerberg, or dark T-shirts like Google cofounder Sergey Brin. I am not saying these people should dress any differently, only that it is easy to believe that the creative is primarily expressed in certain forms. When we believe this, we focus more on fitting a particular physical form and less on accessing the formless energy of the creative.

We can witness clinging to form in a variety of ways. There is a movement, particularly in the tech industry, away from the traditional coat-and-tie corporate style to more casual attire. This is great, but in many cases it is not so much a move toward freedom from form, but simply a replacement of one look with another. Though the latter model may encourage more comfortable clothes, it can be just as limiting. In fact, in some tech businesses, if someone shows up with a coat and tie, people give him no respect. He is seen as an uptight remnant of the old school. People think, *That is not how a smart, creative person should dress. He should look like this other model I have in my head.* The person is discredited because he does not fit others' image of a creative, smart person.

When asked by an interviewer, "How do you systematize innovation?" Steve Jobs answered, "The system is that there is no system."[32]

Whatever the model is does not matter. We can be just as attached to no suit and tie as we can be to wearing one. Both are simply beliefs that try to box the creative into form, which of course only stifles the creative. In this frame of mind, we judge ourselves and others based on this model, and we do not see that the creative can just as easily be expressed by someone wearing a tuxedo as it can by someone wearing grungy jeans and an old T-shirt.

Hold it, you may counter, *if I do not base my decisions on my mental images, how on earth do I make decisions, such as what to wear? I need to wear something. And aren't some forms better than others?*

When the mental images are no longer believed, something else emerges that is more aligned with our inner experience. In fact, our clothes may not change at all, but our attitude will. When we no longer believe there is a look of the creative, we are freer to express whatever form feels right. As such, the next time someone gives a training at our work or school and is dressed differently from our ideas of how a creative person should look (she is wearing a suit or a grungy shirt—whatever counters our idea), we will see her with an open mind, knowing she may well have valuable material to offer. And when someone shows up and fits the image in our mind of how a smart, creative person should look, we will know that the person may or may not have something useful to offer. We will focus on our actual experience instead of how someone does or does not fit our mental image. Of course, some dress codes are needed and there are accepted forms of behavior in any society, but these can exist without boxing the creative.

This also shows up in our ideas of creative activity. If we believe we are creative only through work, then when our partner wants to spend time with us or the trash needs to be taken out, we view these as unworthy of our attention. We think, *How will any of this help me progress further on my project?* In other words, the creative is found in our work, not in these other things. In fact, some of the most unpleasant people to be around are those who identify as "creative people" and view everything other than their work as unimportant and unworthy of their complete attention. This, of course, only produces more tension since the greater their non-work activities, which they translate as non-creative time, the more irritated they become. Not only do they give less attention to other tasks, but when they do finally return to work, they bring this tension and frustration with them.

We may also believe that it is through owning particular gadgets that we are creative. We think we must have the latest iPhone or high-powered Dell computer or most recent software from Adobe in order to do our work. We believe that it is in these forms that we access the creative and that the more sophisticated the technologies we own, the more creative we will be. Though all these tools can be useful, if we do not approach them in the right state of mind, they can actually distance us from the creative mind since we are looking for the creative externally instead of internally. Of course, the problem is not that we own the gadgets, but in believing that they enable us to more easily access the creative mind.

Our ideas of what it takes to live creatively—how we should look, what activities qualify, and what gadgets we need—are simply that:

ideas. In this tool, when we notice ourselves caught in these ideas, we kill the images—or more specifically, we know that these are simply concepts, and when not held to they do not bind or limit us. This is not the end of the arising of such ideas, but the end of believing and identifying with them.

APPLY TO LIFE

The next time you meet someone and find yourself thinking, *He does not look creative,* or you must do an activity that you view as unimportant, thinking, *That is not creativity,* know that these are ideas you hold, that these are not inherent in a person or an activity.

Next, open to the possibility that you can learn from someone who does not fit your idea of a smart, creative person, or that you can make an action creative through the focus you bring to it. See the wisdom in the person who looks like he is not smart, and explore how to creatively take out the trash. Expand your ideas of what it means and doesn't mean to live and be creative. Don't let such ideas bind you.

After doing this, notice if you are able to develop a momentum of creativity that carries over into your other work.

TOOL #24: PRACTICE NON-DOING

Don't Just Do Something, Sit There

To learn how to do, it helps to
first know how to non-do.

To do anything, we must also know its opposite. To code well, we must know how to non-code; to design, how to non-design; to write, how to non-write, to . . . well, you get the point. For example, much of my writing happens in my non-writing time. It arises when walking in nature, doing the dishes, or out on a drive. While I am not writing, writing appears. It's not that I try to write in such times; in fact, it is due to my non-trying that information can flow more smoothly. This is what in some traditions is called non-doing or non-action.

I was taught this by a Japanese businessman who picked me up while I was hitchhiking across Japan. At the time, I was spending four months walking and hitchhiking with friends through the country. We spent several months journeying by foot from Tokyo to Hiroshima as part of a group called the Global Walk for a Livable World, a trek that began on the California coast three and a half years earlier. We also joined a walk for indigenous people on the island of Hokkaido, and explored the country on various side trips. It was great fun. We traveled with no computer or cell phone and with backpacks that weighed no more than twenty pounds, holding one or two changes of clothes, a sleeping bag, and toiletries. Since we had very little money, we often slept in parks and train stations at night. Though most days we walked, when we needed to travel between the two walks or visit Tokyo to deal with our visas or other issues, our most common form of transportation was hitchhiking.

This day, my driver was a wealthy Japanese businessman who was on his way to pick up radio equipment for his newest yacht. Because my destination was near his, he offered to give me a ride. Our conversation soon moved to his software business, which had been quite lucrative. He said that the secret of his success was not what happened during the week at work, but during the weekends on his yacht. On the yacht, he said, he didn't think about or discuss business. He never invited business associates to join him and told friends and family that any talk of work was off limits. His yacht was a work-free zone.

However, he said that it was in this time non-thinking about work that his best ideas for products emerged. Out on the yacht, without trying to think about work, good ideas came. He would then make a note of them and give them more attention during the week. This, he said, was the secret of his financial success.

This was one of my first lessons in non-doing. Of course, both activity and non-activity are needed, both noise and silence, both movement and stillness. However, most of us have been conditioned to worship action. From a young age we are told, "Don't just sit there, do something!" Our *doing*, however, is often not directed or inspired. It is simply action for action's sake. We think if we are doing something—even if it is random, unfocused, uninspired doing—we are being productive. Often, though we are doing a lot, we are getting very little done.

"Do you have the patience to wait till your mud settles and the water is clear? Can you remain unmoving till the right action arises by itself?"

—Chinese philosopher Lao-tze

Another approach is to focus not on action for action's sake but on inspired action. For this we need to not only know how to do, but also how to be, or non-do. Instead of living by, "Don't just sit there, do something," we also know when to follow the commandment, "Don't just do something, sit there!"

In non-doing, it is not a matter of thinking, *Let's see, I could do this or I could do that. If I did this, then this could happen . . .* This kind of thinking has its place, but this is not non-thinking. In non-thinking, we do not follow the various associated thoughts in an attempt to figure something out. Instead, we let go of actively thinking and allow ourselves to receive information. Ideas come more as hits than as carefully thought-out decisions.

The secret to non-doing is to make space for seemingly nothing to happen. Only when the tension around making something happen subsides can something else emerge. Until we can be at peace with nothing happening, in a strange way nothing really can happen since our actions will be an avoidance of non-doing. They will arise from force instead of the natural flow. This is explained by a sculptor in the following ancient story:

> *There once lived a master sculptor in ancient India whose work was so regarded that the king insisted on a visit with him to learn the secret of his skill. When the meeting took place and the king inquired as to the method he used to make such exquisite sculptures, the artist responded, "First, a large slab of stone is brought to me. I then observe the stone. I get to know it, sitting very silently*

sometimes for many days. Then slowly an image starts to emerge. After some time sitting with it in silence, I see what the stone wants to become! It tells me what to do. Only after I have taken the time to know the stone by watching it carefully do I start to carve."

Another way of expressing this lesson is, "Through non-action, skillful action emerges." We no longer take action for action's sake, but instead prioritize inspired or focused action. We see that sometimes the best way to do the most is through non-doing. It is through waiting and watching, and listening to what wants to happen with a project—how a Web site wants to be built, how a product at work wants to be created, how our profile page wants to look.

In this tool, we make room for non-doing, giving time for the right ideas and answers to show themselves.

APPLY TO LIFE

The next time you find yourself working a great deal on a project but not sure what comes next, try this:

- Take time to non-do or non-work. In this time, don't read work-related books, e-mails, or calls. If you find yourself thinking about the issue—trying to figure out a problem with a colleague, mentally composing an e-mail you need to send to an associate, worried about sales this past month—let it go and bring your attention to something else. Tell yourself you can think about it later.

- Then see what emerges in this space. If an idea arises as you non-think, make note of it. Allow thoughts to arise, but don't actively develop any of them until your non-thinking time is over.
- If at the end of this time you find yourself frustrated, thinking, *I have been non-thinking about work all day, and I haven't gotten shit done. I still have not figured out the problem I am having with my colleague.* At such times, you have not been non-thinking. You have been thinking in the guise of non-thinking. True non-thinking has no agenda, no expectations. It is not trying to make anything happen.

TOOL #25: BE SPACIOUS WITH THE GOOD

Get by Letting Go

In trying to hold the creative, we lose it. When, however, we are spacious with the good, the creative more often comes because we allow it to pass.

"It's happening," you say to yourself. The moment of immense creative expression has come. Or in other words, you are in the zone, or flow. Good ideas are emerging effortlessly, blocks are fading away, and you are making connections to elements you had not seen before.

It proceeds like this for some time, and then you think, *This is awesome. How can I keep this going? How can I make this last longer?* No sooner do you have the thought and switch your focus than the zone leaves. The moment you think about how you can hold it, you lose it.

Many situations can cause us to lose our focus and access to the creative mind. It is easy to see how generally unpleasant events do this: someone is rude to us, a project is not proceeding as we had planned, or we have a conflict with a coworker. However, like the previous example, positive events can have the same impact. As the Chinese philosopher Lao-tze warns, "Success is as dangerous as failure." In fact, even moments accessing the creative can make us lose it. The creative does not cause this, per se, but our relationship to it does.

Once we get a creative thought or are in the zone, we have a tendency to act as if it is a terrible emergency or a once-in-a-lifetime event that may never happen again. We then frantically work as fast and as long as we can to milk this creative impulse for everything it's got. We are consumed with a mixture of fear and greed, "This is awesome, but what if this goes away? How can I hold this for as long as possible? Oh no, oh no, oh no."

In this, we treat the creative like a guest who, on arriving, we shove into a chair, tie down, and do everything in our power to prevent from leaving. We are of course happy to see him, but we believe the friendship is best served through force and control. This, however, is no way to treat a guest—or the creative. By doing so, our house becomes a very non-conducive place for visitors. As a result, people (and the creative) eventually stop visiting.

Instead of trying to capture such moments, another approach is to *be spacious with the good*. This involves letting a positive event like a creative thought come, enjoying it, and (here comes the hard part) letting it go, trusting that it will come back in its own time. Often we feel tension around any experience we label as good. This is not inherent in the experience, but arises in our desire to control it, to prevent it from its natural course, which is to arise and pass away.

Recent studies using advanced brain-scanning technologies suggest that true happiness has little to do with our situation in life and much more with how much we have cultivated a sense of inner well-being.[33]

I have seen this dynamic in friends who have acquired large amounts of money either through inheritance or company acquisition. Previously, they worried about money at times, but they lived a pretty free life, knowing that money comes and goes. However, once this "good" event occurred, at times netting them millions, spaciousness was lost, because now they had something to lose. Instead of seeing that money comes and goes, they thought, *I have this money, and who might try to get it from me? How can I never lose it?* Flow was lost and their creative life stagnated. Their life got smaller and smaller as their bank account got larger and larger.

Some people, on the other hand, gain great wealth but do not lose spaciousness. They do not believe that the increase in money proves their worth. They do not take on a new identity because of this or try to find ways to ensure that the money lasts for ten generations. The wealth is simply a tool they have received, which has nothing to do with who they really are, and they open to how best it can be used. They see themselves as a channel for both the receiving and sharing of resources.

An increase in money or another so-called good event, then, does not require that we lose spaciousness. Stifling occurs when we relate through fear and clinging, when we try to grab and hold. An example of this is illustrated, once again, by that rascal Nasruddin:

One day a man noticed Nasruddin sitting outside a bar sobbing.

"What is wrong?" the man asked him.

"It's awful," he said. "A few weeks ago my aunt died and left me two hundred fifty thousand dollars in her will."

"I'm sorry you lost your aunt," the man replied, "but it is nice that she left you so much money."

"You do not know the half of it," Nasruddin countered. "It gets worse. Just today I found out that my uncle died and left me five hundred thousand dollars."

"With all this money, why are you crying?" the man inquired.

Nasruddin responded, "Because I have no more aunts or uncles left who will leave me money."

This is all of us at times. Without spaciousness, good turns to bad very quickly. Nothing satisfies. In this state of mind of grasping, thinking there is never enough, the more good news we receive, the more we suffer. Everything is a problem, and no amount of money, fame, or outward success is ever enough.

When we are spacious with the good, however, instead of fear or greed in these moments, we have gratitude. We appreciate a given moment without needing to control or hold it indefinitely. We relate to these moments with trust instead of fear, with openness instead of greed, with letting go instead of holding. We see that the best way to have something is often to let it go. As such, we become a place where the creative visits more often and stays longer when it does.

APPLY TO LIFE

The next time you feel you are in the zone, welcome it as a good host. Imagine that it does not make you any better or worse because of it. Hold it in appreciation, knowing that it comes and goes.

In doing this, the thought, *I want this to stay,* may still arise. That is fine. Be spacious with that. Allow it to come and go. Focus on appreciation instead of control.

Most of us commute daily somewhere, even if just from our bedroom to our home office or to our living room to watch our children. We move from one place to the next in the morning, and then usually back to the same place at night. Others commute long distances, spending an hour or two in a car, train, or subway each day getting to and from work or school. Still others spend much more of their day in some form of travel.

While commuting, we often have the sense of trying to get somewhere. This is natural since that is why we are commuting. However, when this energy begins to direct our life, we have little patience for anything that gets in our way: every driver of a car that cuts in front of us is an idiot, everything that slows us down even a minute is the cause of stress, every setback (of which there is inevitably something) is immensely stress producing. Indeed, some of the most unpleasant people are those who have become addicted to and are ruled by this energy. They can barely sit still for a minute as this "got to do something, got to get somewhere" orientation runs them crazy. Though their body is in a room, they never really arrive. They are both physically and mentally on the go. When we think about this, what greater curse could we say to someone than, "May you never arrive. May you constantly be trying to get where you are not"?

In its annual Work and Education poll, Gallup reported that the average person spends about 46 minutes commuting each day. If someone works 242 days a year every year between age 18 to 65, that is over 8,700 hours or 363 days of our life we spend commuting.[34]

Rather than silently repeat the mantra, "I'm not there, I've got to get there . . . ," several hours a day while commuting, in this daily practice we explore what it might be like to feel not that we never arrive, but that we always arrive. We might call this present arriving. As we commute, we bring a sense of arriving to our commute. We arrive at our car, at a red light, and at the subway or train station. We are always arriving. This practice turns commuting into something of a meditation. Of course, we can still limit our commuting as much as possible, but when we find ourselves doing so, we make the best use of it.

In doing this, our mind will often counter, *But you haven't arrived. Not only that, but you are late!* Certainly our body may be late, but the real question is, Will our mind be too? Will we show up stressed and late, or relaxed and late? As one teacher put it, "The only true arrival that we are guaranteed of is death." This, we could say, is the commute we all share and are on every moment of every day. What matters, then, are all the moments from now until that time. And every day commuting, at the red light, or on the bus or train, we can be present or frustrated; we can mentally arrive or mentally be absent.

If you want to expand this practice, start with five minutes a day doing the following arriving practice:

1. Find a quiet place to sit, either on a chair or cushion. Take a few deep breaths and relax your body. It is preferable to be out in nature or at least near a plant or other living thing.

2. Next, with your eyes open, invite a quality of arriving to be present. See the nature and other forms around you. Notice the tendency to label these as tree or plant, and instead pay attention to shapes and colors.

3. Next, expand your awareness to other sense doors. Notice sounds, smells, tastes, feelings, thoughts. Rather than trying to get a particular experience, focus on receiving what the moment brings, on arriving in your environment.

4. There is no need for pressure or force. Invite a sense of arriving. If you notice your mind veering off to the past or future, that is fine. Bring attention back to your senses, staying present in your environment. Arrive in the environment around you. Be present to see and receive it.

PART 4

THE SEARCH ENGINE: GO FOR TRUTH

It is easy to enter situations with a particular mental frame. We have ideas of how a meeting or other event will go, but our actual experiences of anything—making a million dollars, creating the next great Web site, finding the perfect partner, or going bankrupt—are quite unlike the ideas we held before they happened. The experience does not match our expectation. Nothing is how we think it is. This is also true about our ideas of ourselves and others.

In this fourth section of the Wisdom 2.0 life, we explore the importance of becoming aware of the frame from which we view the world, of the ideas we hold, and of seeing and aligning ourselves with the truth in any situation.

TOOL #26: NOTICE NEGATIVE FRAMES

Life's Search Engine

The words we enter in life's search engine determine our results. When we enter negative terms, it is no surprise that we generally get related results.

At building 43 of the Googleplex, the central one on the campus, the facility has various ways of showing search activity. On a screen to the left of the reception desk is a running list of terms getting entered into Google's search engine from around the world. It is in almost real time, slowed only by the need to sort out profanity and other inappropriate words. Inside to the right of the entrance, just past security, is a screen with a large globe displaying lights streaming up, indicating the places and quantities of Google searches occurring in the world at that time. Much of Africa is fairly dark, while parts of Europe and the United States are consistently active, though less so at night.

In fact, if we had to pick the most significant change in online activity in recent years, it may very well be the rise in searches. People in the United States, for example, conduct over seven billion every month. Search engines are amazing at what they do. We type in a keyword, click Search, and in a matter of seconds, after scanning data from billions of Web pages, they provide us the most relevant results based on their algorithm. Exactly what criteria are used to determine the top-rated items involves extremely complex algorithms. Google, for example, reportedly uses over one hundred factors in determining what to display. What these are is a part of its secret sauce or what Udi Manber, Google's vice president of Engineering and Search Quality, calls its crown jewels. It is what every competitor wants to beat.

Life, in a way, works similarly but with even more complex and unknown algorithms. It too gives us results based on what we mentally enter, or think. Every day, we attend various events—we show up at work, lunch, a meeting, a party, and online on social networks and chats—and for many of them, we mentally enter a search term. We enter (or think) going into an event, *This is going to be unpleasant,* or, *This is going to be pleasant.* And just like online search terms, these impact the results we get.

It is not so different than the travelers in the following story:

Once a traveler came upon an elderly man seated on a bench outside the entrance to a city. The traveler asked him, "What is this town ahead like?"

The elderly man replied, "First, tell me about the last town you visited."

"Oh, it was awful," replied the traveler. "The people were unfriendly, the food was horrible, and the weather was bad."

"Well," said the man, "this next town will probably be about the same." The traveler then went gloomily on his way.

The next day another traveler walked by the same old man seated on the bench and asked about the village ahead. The elderly man again replied by asking him about the last town he visited.

This time the traveler responded, "Oh, it was lovely. The sunsets were beautiful, the people were fascinating, and there were so many interesting places to visit."

The elderly man then replied, "The town ahead will most likely be the same as your last."

The man went happily on his way.

Of course, the mental frame with which we approach an event does not completely determine the results, but it does have an impact. One way we often get caught is through entering negative terms. We consistently approach events thinking, *This is going to suck,* and again and again, that is what we experience. We then usually conclude, *See? I was right. I thought it would suck and it did.* However, we are often not aware of the terms we entered before it, the mental frame with which we approached it.

We could see this frame as sunglasses or shades that we wear. If we have on red shades, for example, when we look out at the world, we conclude, *The world is red.* And we are right, in a way. If people tell us the world is not red, we will argue, "No, I am telling you the world is red. I'm sure of it." With our red lenses on, that is how the world appears. However, the world is not actually red, of course; it only seems that way due to our red shades, the frame in which we are viewing the world. You might notice this when you are famished and drive down the street looking for a restaurant. While doing so, you only see eating establishments. Gas stations, theaters, dry cleaners, and other non-food-serving places are not seen. You look at the world from the frame, "I need food." This is fine for the moment; however, it is problematic if you continue to wear that lens no matter the situation. Then the frame becomes limiting.

In a study on heart disease, it was found that "the severity of patient hopelessness, as measured by standard psychiatric tests, was proportional to the increase in sickness and death."[35]

When we get caught in a negative frame, we can view most of life as problematic . . . and lo and behold, life responds accordingly. In such a state, we don't just view mild events as terribly disappointing, but life also seems to give us more with which to be disappointed. One hundred invitations are available to a Web 2.0 party, and we are the 101st person to request one. Fifty cars have meters over their time in the parking lot, and the officer writes only one ticket out—for our car. Everyone can easily access wireless at the café—everyone but us. Life never disappoints in giving a wealth of reasons to be disappointed.

You'd think we would happily and eagerly let go of such negative frames. I mean, why hold on to what causes us pain? This, once again, has to do with that delicate subject of our identity, of who we think we are. A part of us often enjoys thinking of ourselves as someone for whom nothing works out or someone who is down on our luck. It gives us a role to play in life. We get satisfaction in having the world reinforce our views, even if what we are right about is our own suffering! The outcome may cause us pain, but hey, at least we are right.

If something good were to happen at the business meeting we thought would suck, we would not see it. The lens from which we view the event does not accept it. How could it? Since our identity is tied to the results, if we were to see more, it would put our entire view of who we are into question! And we, or at least our ego, just cannot let that happen. It feeds off the negative results of the search and uses them to reinforce our view of the world and ourselves. To let go of that, according to the ego, would be death. Misery, to the ego, is better than nonexistence.

A significant shift happens when we become aware of the negative search terms we enter. We then no longer believe, "The world is this way. I am right." Instead, we see that our view is simply our view, and that it need not limit our experience of the event. We are no longer bound by it.

You may be thinking, *But my job really does suck. That is how it is. It is not because I am wearing shades that tell me so. No amount of positive thinking will change the fact that my job sucks.*

This is less about positive thinking and more about awareness. For example, there are a number of ways to hate a job. One way to do so is largely unconsciously. We complain to friends about our boss and coworkers, but do little to try to change the situation. In this approach, we experience constriction and tightening. We carry the pain in our body, blame others, and even if something positive were to happen at work, we would not see it since it is outside our frame. However, when a negative event happens, we conclude, "See? I was right." We get satisfaction every time we find more things to dislike about our work. This subtle satisfaction at difficulty or things not going as wanted reveals that we are caught in a negative frame.

Another way to hate our job is with openness and curiosity. We approach it with the perspective, *I am really unhappy here. Let's see what comes next.* The hate or dislike is present, but so is a quality of learning and investigation. In this, we could say that our dislike is fluid instead of solidifying in our body and mind. We acknowledge what is true but we are not looking for, or getting satisfaction from, events reinforcing our perception.

In this tool, we make conscious the negative terms we enter before an event and by doing so are no longer limited by them.

APPLY TO LIFE

The next time you notice yourself thinking, *Boy, this is going to be awful,* before an upcoming event, notice the frame. No need to judge it or beat yourself up for it. Just notice how the world looks from that particular mind-set.

Then, open around the experience. Become curious. Make room for the event to be other than you think.

Better Than Going Well

Because we get the results we search for, it helps to search not for the good, but for the truth.

The other way we often enter events is by thinking, *This is going to go well,* or by asking God or the universe, "Please make this go well." However, if we are not aware, these too can be limiting and cause us frustration. For example, imagine there is an upcoming event that you feel has to go well. It could be a meeting with a potential customer, a job interview, or a date with someone you like. As the event gets closer, you find yourself silently repeating, "This is going to go well. I know this will go well. Please go well." You don't simply want the event to go well; in your mind, it *has* to go well.

It turns out, the outcome is just as you wished: the customer signs a lengthy contract, you are offered a well-paying job, or the person you dated ends up adoring you. Your hopes come true. However, you soon realize that the customer you were so anxious to get is a pain in the ass to work with; that the new job, though it pays well, is unsatisfying and does not use your skills; or that while you are attracted to the new love interest, the two of you have almost nothing in common. You got what you thought you wanted, but in the end it was not what you truly desired.

On reflection, you realize that the signs were present in the initial meeting with the customer, potential employer, or date, but in your fervor for it to go well, you could not see them.

I experienced this after spending about a year with friends trying to create the next great Web application. For much of that time, I silently repeated, "This is going to be awesome. We're almost there." I kept entering this phrase, pleading with the universe to make the project succeed. However, in my desire for it to go well, I could not see any of the possible ways it could not. As such, I had less ability to see what the project actually needed. All I could see was how it was going well. In this sense, our desire for an event to go well, be it starting a Web company, meeting with our manager, or participating in an online game, can actually limit our ability to see clearly, making it more likely that the event or activity will not go well.

We have another choice, though. What if instead of entering *Please go well* in life's search box, we instead entered *Truth*? What if we put going for truth ahead of things going well? When we can do so, we are not as lured by appealing circumstances: we can see when the high-paying customer needs to be refused, the business we are developing should be folded, the lucrative job offer should be turned down, or the person we are dating is more of a friend than a partner. Our mind, of course, will likely counter at such times, "But this customer has a big account," or, "But you put so much time into the business," or, "But the job pays so well," or, "But she is so beautiful." The truth, however, is the truth, and it cares less about these other matters. Since we will get the truth eventually, we might as well do so sooner rather than later. In going for truth we can also, of course, see when it is right to take the job or to enter the business deal. In fact, we can trust this more because we entered the situation without trying to manipulate the results.

"Every child is an artist. The problem is how to remain an artist once we grow up."

—Pablo Picasso

When we focus on seeing the truth when entering an event, there is less friction in our interactions, less push and pull with life. We can breathe more deeply, see more clearly, and trust our intuition. Of course, we may still have ideas of how an event could turn out and even work toward that possibility, but without force and pressure, without limiting our vision. We can trust a deeper wisdom beyond our ideas of what we think is supposed to happen.

In this tool, we put going for truth ahead of things going well. Since this allows us to see events more clearly, unclouded by our desire for them to go well, we can relate to them more effectively. By not getting caught in the "please go well" lens, we increase the chances that an event will go well.

APPLY TO LIFE

The next time you find yourself thinking, *Please go well; this has to go well,* before an event, notice the frame those mental search terms create. Explore what it may be to go for truth. If it helps, you can silently say, "Go for truth," as you enter. Make the intention to enter the event with open eyes and a willingness to see the truth, no matter what it is.

The Danger of Me

When we only see through the lens of me,
we miss what is true.

Some years ago I was active and had a high ranking on a now-defunct social network. On the site, users could give one another positive or negative points based on whether they thought the person was helpful and had good ideas or not. One's total points were displayed next to his user name. Some people had two thousand points while others only had a few. When new users would join, I expected them to give me, an elder on the site with many points, a certain amount of respect. I figured they should value my opinion more than just any old user. I thought my years on the site and high points should earn me some special treatment from newcomers. When new users did not treat me specially and made comments to me as if I were an average user, I often found myself thinking, *How dare they make such comments to me. How dare they not consider my viewpoint.* I had little problem if they made such comments to others, but if they did so to me, it was unacceptable. We could say that I was not viewing the situation clearly, but was instead only seeing it and responding through what we may call "the lens of me."

You may have also noticed yourself viewing life through this lens. When your boss told you that your work has been shoddy, you thought, *How could she say such a thing to* me? When your project was not going as planned, you thought, *How could this happen to* me? When your computer malfunctioned, you thought, *How could it be* my *computer that has problems?* In these situations, while the event may be difficult, the sense of me is what can cause the most tension and prevent us from seeing clearly. We could call this the danger of me.

I had another example of this some time back when I was teaching a program to incarcerated teens in New York City juvenile halls. One time, standing up in front of twenty-five kids at a facility, I gave what I thought was an excellent talk on the power and importance of awareness. In the stress-reduction class I was offering, I talked eloquently about how we can consciously engage in life. I was on a roll and quite happy with myself because the kids were responding favorably.

After the talk, I walked back over to my chair to take a seat. As I lowered my body to the chair, Mr. Consciousness was not paying attention. I missed my chair completely and fell right on my ass. The twenty-five kids broke out into uncontrollable laughter.

If I had witnessed the same event, I too would have laughed. A man missing his seat after talking for thirty minutes about the importance of awareness—that is funny. But in this situation, it was not funny at all, for it was not just anyone falling on their ass; it was *me*. It was my ass! Because it was *me*, I thought their laughter completely inappropriate. How could they laugh at such a thing? Instead of laughing too, I was appalled, all because it was me.

"I have not failed. I've just found 10,000 ways that won't work."

—Thomas Edison

This is not unlike the example given by the Chinese philosopher Chuang Tzu, who said that if you are out in a boat amidst a good amount of fog, and another boat is on a collision course coming toward you, you will likely yell at the boat, "Watch where you are going. Can't you see I am in front of you? What's your problem?" However, this only occurs when you think someone is in the other boat. If, once the boat gets closer, you realize no one is in the boat, you are no longer angry. How could you be angry at a boat with no one in it? Our belief that someone is in the boat is often what ignites our anger.

In my situation, if I could not be me, I would have had no problem laughing along with the kids. But if I am me, if I think someone is in the boat, then such laughter is unacceptable. Essentially, I am basing my response not on the circumstances present, but on whether it adds or subtracts from what we may call my me-ness. If an event adds to it, then I am all for it—*That is me*, I think. *I am that.* If an event does not support my self-image, then I distance myself from it, and get angry at those who associate me with it. In such a relationship, it is very hard to see what is true.

OK, you may wonder, *but how do I not be me? That is ridiculous, if not impossible.*

As long as we are alive, there will be consciousness in the body. This does not mean, however, that we need to carry a strong sense of *me*—I am this kind of person and not that, I am better or worse than someone else. Of course, we still need to respond to life. If someone is acting like a jerk, we may need to confront the person. But actions can emerge from the needs of the situation instead of an attempt to keep our self-image intact.

This self-image is often supported by the story we tell about ourselves. For example, imagine two people, one man and one woman, each have a million dollars in the bank. The woman, who grew up poor, believes this is a great deal of money and creates a success story around this fact. She believes this is a huge accomplishment and uses it to tell other people what a success she is. She may even write a book or hold seminars on how others can achieve the same success she has had. To her, this million dollars means, "I am a success."

The man, on the other hand, tells a different story based on having that same amount of money in the bank. Because his parents made tens of millions of dollars and he expected to make much more, he tells a failure story based on the million dollars in the bank. He may spend time with people who have a great deal more money, so he interprets the fact that he only has a million dollars as, "I am a failure."

So, does having a million dollars in the bank mean you are a success or a failure? Neither. It depends on the story we want to tell. The million dollars in the bank is a fact. Whether someone then determines that the moneyholder is a success or a failure based on such information is a story. Telling such stories is fine, but there is a world of difference between thinking they are true and knowing they are stories. If we cannot differentiate these, we will always be in conflict with life, trying to get other people to reinforce our story and getting upset when they do not, when they tell us we are a failure when we think of ourselves as a success.

Such stories are different than the functions or roles we play in life as parent, manager, or colleague. It is simply a fact that we have a job as a manager or Web designer. That is one of the roles we play in life. We add me-ness when we believe that such roles make us better or worse than others, that we deserve more or less respect because of them. Likewise, if someone is rude to us, we may very well get upset and need to take appropriate action. The lens of me is present, however, when we do not respond based on someone's actions, but do so because it counters the story we tell about ourselves. We respond out of a desire to uphold and defend our view of who we are. The danger of me surfaces when we find ourselves thinking, *How dare they say that to* me, *considering who I am.*

In this tool, we see that the story we tell of who we are is fine to tell, but it is just a story. Knowing this, we are no longer bound or limited by it. We can then see situations more clearly, unclouded by how they add or subtract from our me-ness.

APPLY TO LIFE

The next time you notice yourself upset because of the me-ness in a situation—"How could they not invite *me* to the meeting?" or "How could she say such a thing to *me*?"—try this:

- Feel the tension present in your mind and body in this perspective, the frustration that people are acting in a way other than how you think they should. Notice the underlying story that is present, the view or views of yourself that you hold. What is the story underlying the sense of superiority or inferiority that you feel?

- Next, expand your view. See the action from outside the story and inquire how you might relate fluidly to it. By fluidly, I mean without holding to the idea of me. If you think you should have been invited to a meeting, see the difference between the functional aspects you think you could provide at such a meeting, if there are any, and the ego or self-serving reasons you wanted to take part.
- Explore a creative response, unhindered by the story. Ask yourself, "If I was responding creatively and fluidly to this moment, how would that look? If my ego and sense of importance were not the primary motivation, how else might I respond?"

When asked what the first thing is he thinks about when he wakes up in the morning, Google CEO Eric Schmidt replied, "I get on my email."[36] When asked again what he thinks about while lying in bed when his eyes first open, he responded, "I open my email." His answer to what he first thinks about is essentially, "I wonder what emails I have received." The other day while watching Justin Kan on the twenty-four-hour-a-day, seven-day-a-week video stream of his life on Justin.tv, I saw him wake up and the first thing he did, even before getting out of bed, was the same: check his e-mail. This is likely true for many of us. It is one of our first daily activities. I do not know Schmidt's or Justin's daily routine (though I guess I could learn Justin's if I had the time and interest), but I do know that how we wake and enter sleep have a big impact on our day.

Through these activities, we essentially have a beginning and ending to every day. And beginnings and endings matter a great deal. If you start reading a novel, for example, but the first fifty pages are not interesting, you will not likely read more. The last two hundred pages could be extraordinary, but without an engaging beginning most people will not read further. Likewise, if the first two hundred pages are great, but the ending is bad, you will likely be dissatisfied with the book. However, if the first fifty pages are good, and it ends well in the last fifty pages, you will be more apt to think highly of the book, even if the middle one hundred fifty pages are just average. In many ways, the same is true with our days. How we start and end a day makes a significant difference in our level of energy and creativity. It sets the tone for our waking and sleeping.

How do we start and end the day? Do we wake up and start our day by turning on our computer to instantly enter the online world to see who may have e-mailed us or what messages we received in a social network, essentially saying to the world, "Tell me what I missed while I was sleeping"? Or do we wake up with an activity that encourages and aids our well-being and creativity, saying to the world, "Here I am. Let me make the most of this day"?

Americans spend over twenty-three billion dollars each year to battle insomnia and other sleep disorders.[37]

Likewise, do we enter sleep only after staying up as late as possible surfing online or watching TV? Or do we turn off technology at a certain time and peacefully make the transition from waking to sleep? Do we view sleep as a nuisance, fighting it for as long as we can, or do we find a relaxing way to depart the day and welcome the night? What is the message we tell the world as we wake and sleep?

In this daily practice, every morning make the first thirty minutes a practice time. Do activities that help you meet the day with calm and clarity. And every night, spend an hour with activities that help you transition from the day to sleep. The activities should focus less on achieving and accomplishing, and more on relaxation and ease, with attention to your inner instead of your outer life. The kinds of practices are up to you, but many people find that meditating, praying, or exercising in the morning greatly enhances their day, and that taking a warm bath, going for a walk, or reading a good book helps them relax at night.

You may be thinking, *Thirty minutes in the morning and an hour at night? There is no way I have that amount of time! My mornings and evenings are packed. I have more important things to do.* Of course, we all have limited time, but if we look more closely, most of us can find the necessary time. For example, can that hour at night we spend surfing the Web or on a social network be cut to thirty minutes? Can those thirty minutes in the morning reading news online be cut to fifteen? Can we play that online game for twenty minutes instead of an hour? Usually we can. Such time on the computer is generally not as vital as we think.

If we take this time, we are not bogged down by work-related items in checking our e-mail five minutes before we go to bed or by waking up and entering the world of activity after little respite. We work enough during the day; do we really want to continue our work in the early morning and before we sleep? The good news is that when we add calming and rejuvenating practices, we often have more energy, need to sleep less, and have a greater ability to focus during our day. In this sense, often such practices in the morning and night actually save us time through less stress, fewer mistakes, and a greater ability to focus. We actually gain more than we lose.

We do not have to spend this hour and a half each day sitting perfectly still, cross-legged, and meditating in a full lotus position. We may set the first thirty minutes in the morning as simply a time of no technology. Instead of instantly checking our e-mail, reading the news, or filling our mind with other content, we let our mind and body gradually adjust to the waking state and take this time to enjoy our breakfast. At night, we could read books or listen to talks that are calming and educational, or we could play with our child or take a walk. The main point is to reduce the pace and intensity of our lives during this time, to have an easy entrance and exit to the day. There is nothing wrong, of course, with looking at a screen or talking on the phone, but if that is how we spend much of our day, it helps to have breaks from such activity, to have non-screen time.

As a trial, for one week do relaxing and rejuvenating activities for the first thirty minutes in the morning and the last hour at night. After the week, you can decide whether or not to continue it. If it helps, you can incorporate some of the following:

1. When waking in the morning, lie still for a few minutes before getting out of bed. Notice the shift from sleeping to waking. Take three full, conscious breaths. Let your system slowly adjust to waking mode. Have a sense of welcoming the day. As the writer

and naturalist Henry David Thoreau said, "Only that day dawns to which we are awake."

2. During both these times, turn or keep off as much technology as possible, particularly communication technologies such computers, TVs, and cells. In the morning, leave your computer and cell off, and turn them off the last hour before going to bed. It can also help to dim lights or, even better, use candles for lighting. This can be particularly helpful the final hour before bed. The softer light gives the mind and body an easier transition to sleep.

3. Avoid potentially contentious conversations. This is generally not the best time to communicate with the sibling, friend, or neighbor with which you have a conflict. Be careful what you put in your mind during this time, either through conversations or various kinds of media. Know that what you consume, you will bring to your day or your sleep.

4. If it helps, make sure others in your household know you will be doing this so they can be supportive.

5. Find and use practices such as meditation, yoga, chi-gong, prayer, walking, running, or relaxation techniques in addition to materials such as music, books, and talks that help to refresh your mind and body. If you want to play a CD or your iPod, do so with talks or music that serve your inner life.

TOOL #29: TRUST NOT KNOWING

What Even Search Engines Don't Know

It is nice to know, but not knowing
is equally as valuable.

Though we are often proud of how much we know, we generally don't know more than we think. This is revealed in the following somewhat playful story:

One morning, a Zen priest was walking from his home across a courtyard when he was stopped by a policeman who was feeling quite combative. "Hey, priest, where are you going?" the policeman asked.

"I don't know," replied the Zen teacher.

"What do you mean you don't know?" said the policeman, angrily. "For the last year, I have watched you walk back and forth from your house to the temple each morning for prayers, and that is where you are going today, isn't it?"

"Maybe. I don't know where I will end up this morning," the Zen priest replied again.

"You smart-ass," replied the policeman. "Do you think I am an idiot? You know where you are going. You are going to the temple."

The Zen priest stood calmly and replied, "Maybe. I don't know."

"OK, wise guy," the policeman announced. "I am taking you in to the police station. Come with me."

In the police car on their way to the station, from the backseat the Zen priest said, "See, I told you I didn't know."

We too never know. In fact, even search engines don't know much. Of course, they are phenomenal at what they do, giving us the most relevant information from indexing billions of Web pages and following user activity. They have created a revolution in the finding and sharing of information with great benefits. However, as cool as they are, they also don't know much. They don't know what will happen to us today, or what we will be doing a year from now, and they certainly do not know what our life's purpose is. This is nothing against search engines, but when we look at all the possible kinds of information we seek, search engines don't know much either.

Our internal search engine equally does not know all the answers. We too don't know much more than we know. The tendency, however, is to attach to our knowing and to hide our not knowing, but doing so makes it harder to see what is true. Imagine, for example, this situation:

The owner of the company you work for invites you to a high-level meeting in which she wants to hear staff's creative ideas for new products. "I want real out-of-the-box ideas," she says to the group. "Give me the best you have."

In the following weeks, you take time to think of creative product ideas. You ask friends, brainstorm possibilities, research other products, but no compelling ideas emerge.

Once at the meeting, the owner goes around the room and has all the staff share what they have come up with the last few weeks. She makes her way through the five people on your left (all of whom have an idea to offer), and then she comes to you.

"What are your ideas?" she asks, and peers directly at you.

In such a situation, how do we respond? If we are uncomfortable with not knowing, then the last few weeks have been hell, and in this moment we are absolutely terrified with thoughts of blame and self-judgment, thinking, *I am such a loser. I don't have any good ideas, and now my boss and everyone else will know. How could I not have one idea? Shit. I better make something up, fast.* Amidst such a state of mind, if there is a creative idea within us, we have almost no chance of accessing it. There is too much mental noise and self-judgment to do so.

"The only thing that interferes with my learning is my education."

—Albert Einstein

However, if we are comfortable with not knowing, then the last few weeks have not been that stressful, and in the moment we are asked by the owner of the company, we take a deep breath, and open to what emerges. It is possible that still no great idea arises at the time and we respond to her, "Nothing has come to me yet." However, it could also be that in that moment of trusting ourselves, of completely letting go and accepting the moment, an amazing idea comes forth from deep within us, something truly out of the box. It arises because we created a fertile field for it to grow, unblocked by fear and self-judgment.

When we can trust not knowing, we are not as propelled by fear (*What if I look like an idiot? What will people think if I don't have a great idea?*) or by expectation (*This is my chance to really impress my boss and colleagues. If I do that, I will surely get a promotion.*). These thoughts may still arise, but we trust the truth of the moment more than our hopes and fears. We trust that if we don't know, then we don't know, and we do not make a problem out of it.

Now, you may be thinking, *This is the worst advice ever. It will surely get me fired. If I am told to bring ideas to a meeting, I better arrive with some.*

I am not suggesting that people always enter meetings unprepared. Put forth whatever effort feels right. However, if we have little trust in not knowing, then the creative, out-of-the-box idea we so desire is harder to access because it is covered by our hopes and fears. As a result, what we do offer comes from tension instead of ease, and does not have the richness and vibrancy we really want.

Of course, at times not knowing can come from laziness or despair. We can say, "I don't know," from an unwillingness to put forth the necessary effort. We can spend weeks goofing off and show up to the meeting with our colleagues, saying we don't know, due to a lack of interest and commitment. We can also say we don't know with our head dropped in embarrassment and shame because of disappointment and self-judgment.

However, yet another way to not know is in a radical openness to life. "I don't know. Far out. How exciting," we say with an open mind, a soft belly, and a deep breath. It is simply the truth of the moment, and we relate to it with ease and spaciousness. There is no problem in it, and as such, ironically, knowing is more likely to emerge.

APPLY TO LIFE

The next time someone asks you a question and you don't know the answer, respond with an expansive don't know. Avoid making a problem of not knowing. See it simply as the truth of the moment.

The Moment Knows

There is no way to know what to say at a future moment because so many conditions are unknown. In that moment, however, we can know.

How many times have you planned exactly what you were going to do or say, followed through perfectly, but not accomplished the results you wanted? It can happen in many situations. Imagine you want to ask your boss for a raise, so you prepare by reading articles online and a book on the subject. From this, you create a precise plan for how and at what stage in your weekly meeting to deliver your request. You have done your homework and are ready to ask your boss for a raise.

The meeting occurs and you make your request just as you had planned, but it feels awkward and forced. It is met with resistance by your boss. "I can't deal with this now," she says. "This is not the time."

What? you think, *I planned it so well and delivered it perfectly. How could this have gone wrong?*

These times can be extremely frustrating since we often delivered just as we had planned. However, the result was not at all what we wanted. Often, there was nothing wrong in our preparation or delivery. The difficulty was that the moment we imagined (in which our planned words may have indeed been just right) was not the moment that occurred. Conditions turned out to be other than we thought. Of course, some ways to express ourselves are better ways than others, whether in asking for a raise or in giving a presentation, but no matter how well we plan our words, much depends on a given moment—and what is most useful is only known in that moment.

"I didn't start out building a search engine. I just said, 'Oh, the links on the Web are probably interesting. Why don't we try doing something with that?'"[38]

—Larry Page, cofounder of Google

This is because no matter how much we imagine a future moment, our experience of it is always other than our vision of it. Life is never how we think it will be. How could it be? At the meeting, your boss could arrive in a number of possible moods, all impacted by countless factors: she could be upset due to an earlier phone call, elated over good news recently received, tired due to sickness, or enthusiastic about some new business model. Who knows what her mind state will be? You too will be influenced by your day, and enter with your own mind state. How can we know ahead of time what the nature of that moment will be while there are so many factors at play? But all is not lost. The good news is that we can know in the moment. Or in other words, the moment knows.

You could be thinking, *Dude, in the last chapter you told us that we don't know more than we know. Now you are telling us that we know in the moment. Explain yourself.*

There is, of course, much that we don't know. We may not know particular information or how to best move a project forward, but we can know the right response in any given situation. In the latter, we may say we don't know, but we do so from knowing that is the best response in that moment. The knowing of the moment is less about particular facts, and more about knowing what is right in a particular moment. This only comes from tuning in to and being present for a given moment.

Without this, we can spend hours imagining how a moment will be, but this effort will be of little use. We end up wasting much of our time, like the travelers in the following story:

One day, two travelers noticed a man standing on a high hill. Wondering what he was doing, one traveler remarked, "He must be standing there because he lost a dog and is looking for it. That has to be the reason."

The other one disagreed. "No, he is standing there because he is looking for land to buy."

The discussion continued for some time, each person arguing a different point. The conversation became quite contentious.

Finally, the two reached the top of the hill. "What are you doing up here?" the travelers asked the man. "Are you standing here because you lost your dog?"

"No," the man replied.

"Are you standing here because you are looking to buy land in the area?"

Again the answer was no.

"Then why are you standing here?" the two travelers asked in unison.

To which the man replied, "I am standing here just to stand here."

In many situations the information does not come as easily as it did to the travelers in the story. In preparing to ask our boss for a raise, she may not say to us, "If you are thinking about asking for a raise at this meeting, now is not a good time to do so, as I am in a very bad mood," or, "The best way to ask me for a raise at the time is . . ." However, we can listen to the moment.

Another more recent example of this is given by Marissa Mayer, vice president of Search Products and User Experience at Google.[39] She tells about a time in creating Google News when the team could not decide which of two features to add: either sort by date or search by location. Team members wondered if people would want to see most recent news or the news closest to their area. The team was at a crossroads, with half arguing for the date feature and the other half for location. The launch date was approaching, so they decided not to include either feature and to see what response they got. The day it launched, Mayer said they got about 305 messages regarding the service. Three hundred of them asked for a sort-by-date feature, and only a few wanted to search by location. Then, they got their answer.

Other times, this listening to the wisdom of the moment takes more work. We do not get three hundred e-mails telling us what to do about a topic of contention. When we are considering asking our boss for a raise, we need to listen deeply to what is most needed at that moment. This requires that we not only pay attention to the words someone uses, but to the spirit of the moment, to *feel* what is right. This requires an intuitive knowing, a deeper kind of listening. The philosopher Chuang Tzu said that hearing with the ears is one thing, and the hearing of understanding is another, but that there is also a "hearing of the spirit." In this, he claims, "When the faculties are empty, then the whole being listens." As we listen with our whole being, we can more easily see the truth of the moment.

Of course, planning can help access possible skillful words and phrases. It's not as if we have to walk in to ask our boss for a raise or give a presentation to a group completely unprepared. We can enter with a very detailed plan and still listen to the moment. From such listening, we may decide to go ahead with our plan, or it could be that we realize in doing so that something else is more appropriate and we adapt as necessary.

In this tool, we see that while planning can be useful, the real answer only comes in a given moment. The moment knows, and it will tell us if we listen.

APPLY TO LIFE

The next time you are wondering what you will do or say at a future event, try this:

- Notice the level of fear or trust that is present. Inquire as to what is motivating you. Ask yourself, "If I related to this event with a deep trust of the moment, how would I act?" If planning seems important, do so. But know that whatever you decide to say beforehand is simply one possibility and that it may or may not be useful once the time comes.

- When the moment does arise, feel what is present. Open all your senses to take in the moment. As Chuang Tzu suggests, listen with your whole being. If it helps, silently ask, "What is the truth of this moment? What wants to come forth?" Stay open to what the moment knows.

Be Right and Wrong

It is nice to be right, but it is even better to be in touch with truth, no matter whose mouth it comes from.

What if someone asked you, "What are the top things you love the most?" You may think of your parents, your partner, a dear friend, or—if you are a total techie—your beloved MacAir or iPhone. However, while all these may be true, there is another one that we often miss: our thoughts. Though on the surface this may seem like an outrageous statement, what has waged almost every war in history and caused almost every argument either in person or online? Thoughts. Of course, thoughts themselves are not causing the arguments, but our love of and attachment to our thoughts, our need to defend and prove how much our thoughts are better than others'.

There are certainly worthy battles to fight and arguments to have. However, when we attach to our thoughts, when we have a strong need to be seen as right, we can easily miss the truth. I was caught up in this some time ago. I woke up at two in the morning, unable to sleep after a heated online debate earlier that day about the pros and cons of Wal-Mart. The other person did not see the issue my way, and that afternoon I had written what I thought was a brilliant, informed, completely convincing response to him. If there was a national award for posts online, this would have won it, hands down. With this, I thought, the person would crumble and meekly admit defeat, showing once and for all that I was indeed right.

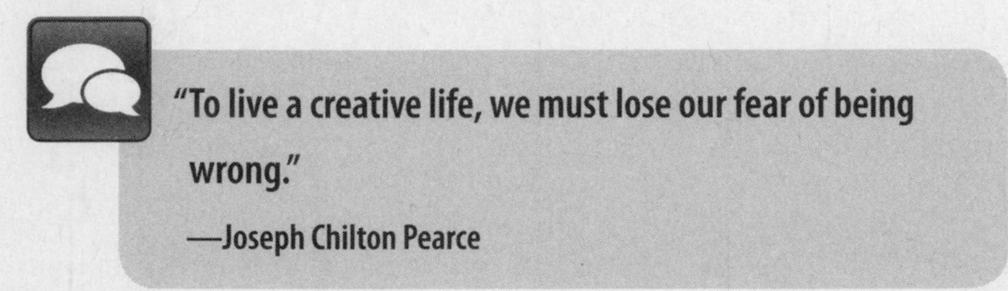

I had waited all night for his concession post, which I assumed would highlight the eloquence and perfection of my comment and reasoning. Though I had spent much of the night checking and refreshing the page, he had not responded. *Could he actually be taking this time to compose a post disagreeing with mine? On what grounds? Surely, I am right.* Finally at midnight, I went to sleep.

At two a.m., I woke up amidst this mental activity, thinking, *Ah, surely by now he has replied. Let's see what he said*. I walked to my desk and waited what seemed like an eternity for my computer to start and for my browser to launch. The page slowly opened. The update: while others had posted comments, the person I was waiting for had not. I went to his profile page and saw that he lived on the east coast, a couple hours ahead of me. I calculated when he would likely wake up and decided I would try again at about seven in the morning. Still, I was perplexed. What on earth was taking him so long?

Then it hit me: I was up at two in the morning unable to sleep, consumed by a debate with a person I had never met, on a subject I only partially cared about, and on a site that gets thousands of posts a day. In this, I had one of those moments, all at once terrifying and enlightening: I was immersed in this discussion in large part because, well, I wanted to be right and I would not rest until I was acknowledged as such.

Who was really right in the discussion I was having? The answer very well could be much like that in the following Zen story:

A student asked a Zen teacher, "To get enlightened, is it true that the most important thing is simply to surrender and trust life, that no work is needed?"

"You are right," replied the teacher.

"But," replied another student, "haven't you said in the past that the most important thing is to prioritize awakening and that one must work hard to realize it?"

"Yes, you are right."

"Hold it," countered a third student. "There is no way that both of them can be right since the approaches oppose one another. Either one person is right or the other is. They can't both be."

The Zen teacher paused for a moment, and replied, "You are right, too."

In many conversations, there is no right or wrong, no winner and loser. There are just different views expressed. While sharing our thoughts is important, when we attach to them, when we are sure we are right, we invite conflict. In fact, the stronger we hold to our view, the more we invite others to respond with equal intensity. The group then gets caught in a thought battle, to which there is rarely a winner. In fact, few things can more disrupt a group, limit fresh thinking, and cause turmoil. If you have five people working together who all want to be right, there is little hope for creativity and teamwork.

When we are in this frame of mind, we do not discuss as much as we try to convince. We do not listen, but instead spend the time someone else is talking by planning our rebuttal, which often relates little to his last comment because we did not hear it. In fact, in this frame of mind, we don't really care what someone else thinks; we know we are right and believe that our job is simply to get others to see it our way—and the sooner the better.

Of course, sometimes we may be right about particular facts or have the right idea or answer in a given situation. The question, however, is not only *if* we are right, but *how* we are right. Are we right with spaciousness or with contraction, with ease or with tension? Do we think moments when we are right prove our superiority and those when we are wrong prove our inadequacy? Or can we see that they are both just moments which, if we can take them with equanimity, do not determine our worth?

When we can be right and wrong with equal equanimity, we are not as swayed by either position. We see that rightness is not all it is cracked up to be. It may give us a quick ego rush, but it does not provide real satisfaction. And we see that the times we are wrong are not such devastating events; they happen and pass fairly quickly. In this, the lure to prove ourselves better than someone else and the fear of being seen as less than another lessens. In doing so, when a great idea emerges from someone else's mind, we are more likely to meet it with joy instead of frustration, with ease instead of stress.

We have the ability to discuss differing views with openness, since there is less tension around our ideas and those of others. From doing so, we can see more clearly the most helpful idea or plan in a given situation, no matter who offers it. We may still love our own thoughts, but we love the truth even more. This, we could say, is the real right.

APPLY TO LIFE

The next time you meet with friends or colleagues to discuss ideas, either in person or online, see each idea with equal weight, as if they all arise from the same group mind. You are not any more excited for your idea to be picked than anyone else's. Discuss them without attempting to argue and fight for the idea you have.

Still express your view, but know that both rightness and wrongness are just labels, and each can be experienced with equal ease and spaciousness.

We have all likely had those times when, after leaving a voicemail message or sending an e-mail, we thought, "Shit. I should have never expressed that. Why didn't I think first before I communicated that?" We try to figure out a way to delete the message, but usually it's too late—the e-mail has already been sent or the voicemail left. All we can do is cringe and wait for the person's response. We think to ourselves, *If only I had taken more time before sending the message* . . . In such situations, we have usually created a mess that we later have to clean up, all of which could have been avoided had we only taken a minute (or even fifteen seconds) to settle our mind and body before we expressed ourselves. That one minute could have saved us hours of difficulty.

It is amazing how much communication takes place with almost no mental preparation, as if we view our mental state as having little or no impact on our communication. We rush into our office and quickly pick up our phone or get on Skype, not realizing that we are entering a call agitated. Or we immediately respond to an e-mail before we better understand the frustration we feel from reading it. Our response, then, is not as clear or effective because we are reacting instead of consciously responding.

One valuable daily practice, particularly for those of us who communicate a great deal, is conscious communication. This can be helpful on a number of fronts: First, as in the example above, it can save us time and difficulty as there are fewer messes that we will need to clean up from speaking unskillfully. Second, when our mind is focused and clear we can often say in five minutes what it might otherwise take us to communicate in thirty. And third, it can give us energy by reducing the amount we expend in communication.

Warren Buffet gave Bill and Melinda Gates only one piece of advice on how to spend the over thirty billion dollars he is donating to their foundation: "Stay focused."[40]

In this daily practice, take the time at least once a day to consciously communicate. You can sit in silence for a minute or take three full, conscious breaths before making a call, entering a meeting, or writing an important e-mail. Whatever you decide, the important piece is to give attention to your mental space and to only communicate when you feel at ease in your mind and heart.

The Zen teacher Thich Nhat Hanh encourages what he has coined "telephone meditation." This involves using the ring of the phone as a wake-up call. In this, you do not immediately answer your phone on the first ring, but instead take a few conscious breaths. You only pick up the phone after the third ring. Instead of rushing to answer it or trying to get as much done in the few seconds between these rings, you pay attention and cultivate a healthy mental space before answering. He also suggests taking a few breaths before making a call so you can use words more effectively.

Telephone meditation is not for everyone, but the important element is to give attention to the mental space from which we are communicating, and to see the value in speaking clearly. We do not see communicating as, *If I say or type enough words, eventually something useful will come out*. This may be true, but it is terribly tiresome. Some conversations, of course, are mainly for entertainment and our mental space matters less, but in others it is essential that we engage with clarity, or we will feel the negative consequences.

In this daily practice, once a day make the intention to communicate consciously, taking three full, conscious breaths when needed to calm and focus the mind. Know the power of words. As the Buddha encouraged, "Better than a thousand words mindlessly spoken is one word of truth that helps bring peace to the listener."

If it helps, follow these steps before a call or other communication with which you want to perform consciously:

1. Before the call takes place, take a few minutes to breathe deeply. Let your breath be full and at ease, and soften any tension in your body.

2. As you do, notice any sense of fear or expectation that may be present. Do you worry that if this does not go well, there will be many negative consequences? Or do you hope you will gain a great deal from the call? Do you think you will either lose or gain financial, social, or other benefits? Allow your fears or expectations to be there. Make them conscious.
3. Then, as you keep breathing deeply, drop to a deeper place, one of trust and clarity, a place that knows that beyond the outcome of the conversation, you will still be the same person. Find the place that knows nothing of real importance ever changes. Let the rhythm and flow of your breath guide you to this place. The conversation still matters, but it will not impact your well-being at the deepest level.
4. When you do make the call, keep breathing deeply, trusting that place of clarity. If you notice yourself getting caught in fear or expectation, trying to force the conversation in one direction, come back to your breath.
5. Bring your full attention to the call, practicing both speaking clearly and listening deeply, making sure you are not just understood, but that you also understand the other. Know that if you can bring your full attention to the call, the results will work themselves out.

Afterword

For most of us, there is no question that we will continue to use and benefit from the great technologies of our age and live what we may call 2.0 lives. The question, instead, is whether we will live Stress 2.0 or Addiction 2.0 or Wisdom 2.0, whether we will relate to these devices creatively or stressfully. Those of us fortunate to have access to the Internet and to cell phones will need to answer this question for ourselves. For those of us who walk the Wisdom 2.0 path, the challenge is to bring consciousness into our actions. The Indian philosopher Jiddu Krishnamurti once wrote, "The man who knows how to split the atom but has no love in his heart becomes a monster."[41] In the same way, the person who can be connected all the time through technology, but has no consciousness, no love in his heart and no access to the creative, may not be a monster but will miss the deeper potential and wisdom in life.

The negative effects of actions motivated by fear, greed, and hatred are felt both by us individually and by our earth as a whole. Though our everyday choices may seem small and insignificant, every time we bring consciousness to our actions, every time we engage creatively instead of stressfully, we are helping to shift this momentum. What we will create from such actions—a product at work or a Web site in our spare time—will be an expression of this consciousness.

In doing so, instead of a lose-lose-lose life with more stress, more unfocused work, and more negative impact in the world through a largely unconscious life, we can discover a win-win-win path with more ease, more effectiveness, while making a more positive impact in the world. This is our challenge and our opportunity.

Acknowledgments

I offer a deep thanks to Jack Kornfield, who inadvertently gave me the idea for this book during a phone discussion we had about literary agents. The next day, I stopped work on my other book and started this one.

What insights are present in this book are in large part due to teachers I have learned from over the years—both those who have passed on, such as the Buddha, Rumi, Kabir, and Lao-tze, and modern ones including Stephen and Ondrea Levine, Sharon Salzberg, Jack Kornfield, Ram Dass, Jon Kabat-Zinn, Arnold Mindell, and Joan Halifax. And most recently to the power of the presence of Eckhart Tolle, whose timeless teachings offer such a clear path to a sane, livable world.

For reviewing the manuscript and offering other writing support during the process, I send blessings to Paul Zelizer, Mark Grimes, Judy Buffaloe, Jenna Buffaloe, Patricia Savitri Burbank, Michelle Goguen, Keith Katchick, Will Kabat-Zinn, Heather Harell, Alison Zelizer, Katherine Leiner, and Makenna Goodman.

This work has greatly benefited from conversations with my brother, Brett, who has been a constant guide and friend; my sisters

Zoe, Sola, and Kerena and their significant others (Derek Gerlach, Mark Millard, and Ben Saltzman) have helped through dialogue, reviewing chapters, but most importantly love and support. Thanks as well to my mother, Ann Moore, and father, Rolf Gordhamer, who always give their full support to my efforts.

I offer a deep bow to my agent, Sarah Jane Freymann, who is not only an exceptional literary agent, but also a very kind, insightful, and generous person. I cannot imagine someone better at what she does.

I send my gratitude to my editor, Eric Brandt, for believing in the need for such a book and giving the manuscript his keen eye, and to the team at HarperOne, including Carolyn Allison Holland, who kept the project together; Ali McCart, for skilled fine-tuning; and Ralph Fowler, for the design.

Notes

1. When I use the word *technologies* in this book, I am primarily referring to communication devices and networks, such as cell phones and the Internet.

2. Miniwatts Marketing Group, "Internet Usage Statistics," *Internet World Stats*, http://www.internetworldstats.com/stats.htm.

3. Tarmo Virki, "Global Cellphone Penetration Reaches 50 pct," *Reuters*, November 29, 2007, http://investing.reuters.co.uk/news/articleinvesting.aspx?type=media&storyID=nL29172095.

4. U.S. Department of Health and Human Services, "Stress . . . at Work," *National Institute for Occupational Safety and Health*, http://www.cdc.gov/niosh/pdfs/stress.pdf.

5. Paul Waddington, "Dying for Information? A Report on the Effects of Information Overload in the UK and Worldwide," in *Beyond the Beginning: The Global Digital Library*, report prepared by the Marc Fresko Consultancy for the Coalition for Networked Information, http://www.cni.org/regconfs/1997/ukoln-content/repor~13.html.

6. Matt Richtel, "Lost in E-Mail, Tech Firms Face Self-Made Beast," *New York Times*, June 14, 2008, http://www.nytimes.com/2008/06/14/technology/14email.html?_r=2&pagewanted=1&ref=business.

7. Matt Richtel, "In Web World of 24/7 Stress, Writers Blog Till They Drop," *New York Times*, April 6, 2008, http://www.nytimes.com/2008/04/06/technology/06sweat.html?pagewanted=print.

8. Marc Ransford, "Average Person Spends More Time Using Media Than Anything Else," *Ball State University*, September 23, 2005, http://www.bsu.edu/news/article/0,1370,--36658,00.html.

9. Claudia Wallis, "The Multitasking Generation," *Time*, March 19, 2006, http://www.time.com/time/magazine/article/0,9171,1174696-1,00.html.

10. Ashley McRaven, "Better Breathing for Better Scores," *Southern Illinois University*, August 30, 2007, http://media.www.siude.com/media/storage/paper1096/news/2007/08/30/Campus/Better.Breathing.For.Better.Scores-2942445.shtml.

11. The Gallup Organization, "Attitudes in the American Workplace VI." Poll conducted for the Marlin Company, September 4, 2000.

12. Benedict Carey, "Too Much Stress May Give Genes Gray Hair," *New York Times*, November 30, 2004, http://www.nytimes.com/2004/11/30/health/30age.html?position=&adxnnl=1&fta=y&pagewanted=all&adxnnlx=1213648399-jk/A7tWM8sdkGGEzEuPpNw.

13. American Psychological Association, "Stress in America," October 24, 2007, apahelpcenter.mediaroom.com/file.php/138/Stress+in+America+REPORT+FINAL.doc.

14. Phil Jackson, *Sacred Hoops: Spiritual Lessons of a Hardwood Warrior* (New York: Hyperion, 1996), 4.

15. "Are PDA's Replacing Pillow Talk," a study conducted by Studylogic, sponsored by Sheraton Hotels and Resorts. Read online at: http://www.starwoodhotels.com/sheraton/about/news/news_release_detail.html?obj_id=0900c7b9809c404f

16. American Psychological Association, "Stress in America," October 24, 2007, apahelpcenter.mediaroom.com/file.php/138/Stress+in+America+REPORT+FINAL.doc.

17. Lev Grossman, "The Off-Line American," *Time*, August 14, 2008, http://www.time.com/time/magazine/article/0,9171,1832862,00.html.

18. Paul Taylor, Cary Funk, and Peyton Craighill, "Who's Feeling Rushed? (Hint: Ask a Working Mom)," *Pew Research Center*, February 28, 2006, http://pewsocialtrends.org/assets/pdf/Rushed.pdf.

19. Emily Keller, "Why You Can't Get Any Work Done," *BusinessWeek*, July 19, 2007, http://www.businessweek.com/careers/content/jul2007/ca20070719_880333.htm.

20. Justin Kruger et al., "Egocentrism over E-Mail: Can We Communicate as Well as We Think?" *American Psychological Association*, December 2005, http://psycnet.apa.org/index.cfm?fa=buy.optionToBuy&id=2005–16185–007.

21. Tina Rosenberg, "When Is a Pain Doctor a Drug Pusher?" *New York Times*, June 17, 2007, http://www.nytimes.com/2007/06/17/magazine/17pain-t.html.

22. Phil Jackson, *Sacred Hoops: Spiritual Lessons of a Hardwood Warrior* (New York: Hyperion, 1996), 5.

23. Nikola Tesla, *My Inventions: The Autobiography of Nikola Tesla* (Sioux Falls, SD, 2007), 13.

24. Jessica Livingston, *Founders at Work: Stories of Startups' Early Days* (Berkeley: Apress, 2007), 36.

25. Jessica Livingston, *Founders at Work: Stories of Startups' Early Days* (Berkeley: Apress, 2007), 124.

26. You can find a downloadable mindfulness clock at http://www.mindfulnessdc.org/mindfulclock.html.

27. Medical News Today, "Low-Level Stress Reduced by Nature, Not Technology," June 12, 2008, http://www.medicalnewstoday.com/articles/110799.php.

28. Adam Lashinsky, "Working in the Googleplex," January 22, 2007, http://money.cnn.com/galleries/2007/fortune/0701/gallery.Googleplex_campus/6.html.

29. Larry Brilliant, interview by Daniel Goleman (podcast), *More Than Sound Productions*, January 21, 2008, http://morethansound.net/wordpress/?p=30.

30. Sergey Brin, interview by John Battelle (podcast), *ITConversations*, October 6, 2005, http://itc.conversationsnetwork.org/shows/detail795.html.

31. Craig Newmark, interview by Tavis Smiley, *PBS*, January 23, 2006, http://www.pbs.org/kcet/tavissmiley/archive/200601/20060123_newmark.html.

32. Peter Burrows, "The Seed of Apple's Innovation," *BusinessWeek*, October 12, 2004, http://www.businessweek.com/bwdaily/dnflash/oct2004/nf20041012_4018_db083.htm.

33. Michelle Conlin, "Meditation," *BusinessWeek*, August 30, 2004, http://www.businessweek.com/magazine/content/04_35/b3897439.htm. Books and articles by Daniel Goleman and Richard Davidson are also very helpful.

34. Joseph Carroll, "Workers' Average Commute Round-Trip Is 46 Minutes in a Typical Day," *Gallup News Service*, August 24, 2007, http://www.gallup.com/poll/28504/Workers-Average-Commute-RoundTrip-Minutes-Typical-Day.aspx.

35. Philip J. Hilts, "Pessimism Is Hazardous to Health, a Study Says," *New York Times*, November 25, 1995, http://query.nytimes.com/gst/fullpage.html?sec=health&res=9C06E0DE1139F93AA15752C1A963958260.

36. Eric Schmidt, interview by John Battelle, *The Google Podium*, April 17, 2007, http://scholar.google.com/press/podium/web_expo_2007.html.

37. John LaRosa, "U.S. Sleep Aids Market Grows to $23 Billion, as Americans Battle Insomnia, Sleep Disorders," *PRWeb*, June 9, 2008, http://www.prwebdirect.com/releases/2008/6/prweb1006354.htm.

38. Academy of Achievement, "Sergey Brin & Larry Page," October 28, 2000, http://www.achievement.org/autodoc/page/pag0int-1.

39. Marissa Mayer, *Entrepreneurial Thought Leader Speaker Series* (MPEG), 4 min., 5 sec.; from Stanford University's Entrepreneurship Corner, http://edcorner.stanford.edu/authorMaterialInfo.html?mid=1528.

40. Patricia Sellers, "Melinda Gates Goes Public," *Fortune*, January 7, 2008, http://money.cnn.com/2008/01/04/news/newsmakers/gates.fortune/index3.htm.

41. J. Krishnamurti, *Education and the Significance of Life* (San Francisco: HarperOne, 1981), 19.

About the Author

SOREN GORDHAMER works with individuals and groups on ways to live with less stress and more effectiveness in our technology-rich lives. He has taught classes on stress reduction to such diverse populations as youth in New York City juvenile halls, trauma workers in Rwanda, and to staff at Google's corporate headquarters. The founder of the New York City non-profit Lineage Project, which offers programs to incarcerated teens, Soren has also been involved in creating online tools to help uplift user-generated content. A former project director for Richard Gere's public charity, Healing the Divide, he organized the "Healing through Great Difficulty Conference" with His Holiness the Dalai Lama. Soren is also the author of the meditation book, *Just Say Om!* and has been featured in *GQ*, *Newsweek.com*, and other publications. For more information, visit his Web site at www.sorengordhamer.com.